Ricciocarpus natans

Liverworts
and
Hornworts

of
Southern Michigan

Howard Crum

THE UNIVERSITY OF MICHIGAN HERBARIUM
ANN ARBOR

Most of the costs of manufacturing this book were borne by the
Clarence R. and Florence N. Hanes Fund. Supplementary support
came from The Gillette Fund, administered by the University of
Michigan.

Library of Congress Cataloging-in-Publication Data

Crum, Howard Alvin, 1922–
    Liverworts and hornworts of southern Michigan / Howard
Crum.
        p.   cm.
    Includes bibliographical references and index.
    ISBN 0–9620733–1-8
    1. Liverworts—Michigan—Classification.   2. Hornworts
(Bryophytes)—Michigan—Classification.   3. Liverworts—
Michigan—Identification.   4. Hornworts (Bryophytes)—
Michigan—Identification.   I. Title.
QK556.5.M5C78   1991
588'.3'09774—dc20   91-7992
                                                                CIP

Edited by Christiane Anderson

This book on Michigan's liverworts and hornworts
is dedicated to Michigan's own Bill Steere (1907–89),
talented bryologist, able administrator, good teacher,
pleasant companion, loyal friend.

Contents

Introduction

> I like to look on plants as sentient beings, which live and enjoy their
> lives—which beautify the earth during life, and after death may adorn
> my herbarium. . . . It is true that the Hepaticae have hardly as yet
> yielded any substance to man capable of stupefying him or forcing his
> stomach to empty its contents, nor are they good for food; but if man
> cannot torture them to his uses or abuses . . . they are, at the least,
> useful to, and beautiful in themselves—surely the primary motive for
> every individual existence.
>
> —Richard Spruce

Yes, they are beautiful and, as Spruce said, useful to themselves, but
are mosses and liverworts "good for anything" —of any use to man?
Cotton Mather taught us that anything useless is vicious. Maybe so,
but William Shakespeare said, about the same time, all the uses of
the world are "weary, stale, flat, and unprofitable." And William
Campbell Steere, who was my teacher at the University of Michigan
and later Director of the New York Botanical Garden, told the *New
Yorker* that bryophytes had been useful in keeping him and his family going for 35 years! They have served me and my family well and
have provided me, though not my family, with a never ending joy in
seeking, finding, knowing.

Steere's pleasant little book, *Liverworts of Southern Michigan*,
helped me along in my first try at naming liverworts nearly five
decades past. That book is now out of print and out of date. In
putting together this replacement guide to the liverworts and
hornworts of Michigan's southern counties, I have come to a better
understanding and greater enjoyment of plants I had long neglected
in favor of the mosses. I hope my late-found shock of pleasure can be
passed along.

Steere's book is non-technical, intended only as a primer. Mine is
more informative and, no doubt, less charming. Most of Michigan's
liverwort and hornwort collections have been made by botanists or
students of botany, and I expect that discoveries will continue to be
made by persons with some botanical knowledge and a willingness
to use a microscope. However, most liverworts, small as they are,

1

can be recognized without a microscope, with the unaided eye or with a handlens, and so this guide, with its abundance of illustrations, should not be wholly useless to persons less informed and equipped. The illustrations are intended to supplement or replace descriptive wordiness. Thomas Gray could speak of "thoughts that breathe and words that burn," but as a word-using taxonomist I agree with Victor Schiffner that no language is rich enough to describe 8000 leaf shapes. Scattered in the text is a fair amount of bookish detail on structure and development not essential for identifications but useful for understanding and, I hope, not too wordy. It is the old-fashioned morphology that gives a student something more than a Poll Parrot level of understanding.

The world's hepatic flora appears to consist of some 5500 to 8500 or even 10,000 species, depending on who says it and when. Rudolf Schuster, who has had more experience with liverworts worldwide than anyone else, past or present, has set the total at 5500 or 6000, but he has also said that the leafy liverworts alone may add up to somewhat more than 7000 species. The liverworts of Michigan number perhaps 161 species (Hollensen, 1984). Steere included 58 of them in his *Liverworts of Southern Michigan*, but he failed to define the area under consideration. In the present work, 76 species are recorded from that part of Michigan lying south of the "Tension Zone," where south becomes north in climate, soil, and plant cover [fig. 1].

> How to get to fairyland, by what road, I did not know, nor could anyone inform me.
>
> —Herman Melville

About halfway up the Lower Peninsula of Michigan, anyone with eyes begins to notice a change in vegetation. Farmland is left behind, and the vegetation takes on a different look, owing to an increased abundance of conifers, aspens, and white flashes of paper birch. The North has begun. The upland deciduous forests of oak-hickory and beech-sugar maple give over to a northern hardwood complex of beech, maple, hemlock, yellow birch, and white pine on better upland soils and bigtooth aspen, red maple, red oak, and pines on sandy dryness. Cone-bearing black spruce and white spruce, tamarack, fir, and white cedar rather than broad leaf species—hardwoods—

characterize the wet lowlands. Paper birches are seen in both upland and lowlands, often in places burned over. The change to a more northerly vegetation seems abrupt to a motorized observer, but it actually takes place in a beltlike area of transition—a "Tension Zone" —some 20 to 60 miles wide and extending across the state from Saginaw Bay westward.

I have chosen to consider the liverwort flora of 42 counties lying wholly or partly south of the Tension Zone [fig.1], in a naturally defined area corresponding to "Southern Lower Michigan" as defined by Albert, Denton & Barnes (1986):

Allegan	Huron	Montcalm
Barry	Ingham	Muskegon
Bay	Ionia	Newaygo
Berrien	Isabella	Oakland
Branch	Jackson	Ottawa
Calhoun	Kalamazoo	Saginaw
Cass	Kent	Sanilac
Clare	Lapeer	Shiawassee
Clinton	Lenawee	St. Clair
Eaton	Livingston	St. Joseph
Genesee	Macomb	Tuscola
Gladwin	Mecosta	Van Buren
Gratiot	Midland	Washtenaw
Hillsdale	Monroe	Wayne

Michigan has never really had a present moment. It has a mysterious past and an incalculable future, attractive and terrifying by turns, but the moment where the two meet is always a time of transition. The state is caught between yesterday and tomorrow, existing less for itself than for what it leads to; it is a road whose ends are distorted by imagination and imperfect knowledge. The great American feeling of being en route—to the unknown, to something new, to the fantastic reality that must lie beyond the mists—is perfectly represented here.
—Bruce Catton

Ours is a pleasant state, green and placid, yet scarred by disturbance and devastation, past and present. The gouging, grinding, scraping, flooding, chilling violence of the Pleistocene displaced and impoverished Michigan's flora time and again, and each time hardy survivors re-invaded as the land was bared of ice and meltwater, as climates ameliorated, vegetation changed, and soils matured. Glacial retreat began in Michigan only about 15,000 years ago. The forests that reached maturity since that time have, in the last two centuries, been sadly altered as man slashed and burned, exploited the soils, increased run-off and erosion, silted streams, and polluted soil, water, and air. The forests that remain, however green, are second-growth and scattered, and a reduced flora surviving in jeopardy continues to change.

In spite of progress and pollution, death and destruction, the two peninsulas of Michigan offer a diversity of climate, topography, and soil, as well as vegetation. The Upper Peninsula is rugged and rocky, whereas the less elevated Lower Peninsula has scarcely any outcropping rock. The Lower Peninsula shows some striking changes in climate, day length, soil, and vegetation on either side of the Tension Zone. While many common plants widely distributed in eastern North America range throughout the state, many others rarely, if ever, grow north of the Tension Zone, and some are rarely or never seen south of it.

The Tension Zone separates the fine-grained, gray-brown soils of the southern deciduous forests from sandy, leached soils associated with the acid litter of northern conifers. The south has a more continental climate of extremes, less influenced in temperature and precipitation by the Great Lakes. The growing season is longer, and the danger of a late spring freeze is less. Summer temperatures are higher, and nights are warmer. The precipitation-evaporation balance is less favorable, and summer drought is a certainty. Winter precipitation is greater, but the amount of snowfall is less. The contrast in growing conditions on either side of the Tension Zone (derived from Albert et al., 1986) provides the rationale for treating the liverwort flora of southern Michigan separately from that of the south (Table 1).

Southern Lower Michigan consists mainly of glacial moraines and outwash and post-glacial lake plains. The soils are mainly clays, loams, and outwash sands overlain to a greater or lesser extent by woodland humus or lowland peat and muck. The mineral soil is derived from transported glacial materials and underlying limestone and shale. Finely ground calcareous parent materials and warm summers have favored a relative luxuriance of plant growth. In Northern Lower Michigan, by contrast, a high plain dominates the inland

TABLE 1

	Southern Lower Michigan	Northern Lower Michigan
Length of growing season	146–163 days	115–133 days
Ratio of night heat sum to total heat sum	28–30%	23–28%
Annual average temperature	8.6–9.4°C	6.2–7.8°C
Average temperature May–Sept.	17.8–19.3°C	15.9–17.8°C
Extreme minimum temperature	−22°−−24°C	−22°−−29°C
Potential evapotranspiration May–Sept.	520–540 mm	470–500 mm
Ratio of precipitation to potential evapotranspiration July & Aug.	61–71%	64–70%
Annual precipitation	740–900 mm	740–850 mm
Precipitation May–Sept.	360–440 mm	380–400 mm

where common features are moraines, ridges of ice-contact material, and outwash. The soils are mainly sandy and acid. In a few coastal areas are limestone outcrops.

In southern Michigan, representatives from the oak-hickory forests of Indiana and Ohio contribute to a floristic diversity that is lacking in the north. The "primeval," presettlement forest included stands of beech and sugar maple on loam, oak and hickory on sand. But post-settlement intrusions of bigtooth and trembling aspen, black cherry, and sassafras stand in evidence of disturbance. Hemlock and white pine, once common in the northern part of the lake plain bordering Lake Huron, are now greatly reduced in abundance. Swampy lowlands, or such as escaped drainage, are mainly occupied by deciduous hardwoods, tamarack, or buttonbush. Mineral-rich, sedgy fens are often seen in lowlands, especially at lake margins where water charged with calcium bicarbonate is constantly available.

In Northern Lower Michigan, the original forests consisted of a patchwork of beech, sugar maple, hemlock, yellow birch, and white pine on moist upland soils and pine or oak-pine mixtures on dry soils. The abundance of beech trees fed vast flocks of passenger pigeons until logging and burning slash replaced the original woodland complex with pioneering paper birch, bigtooth aspen, red maple, and red oak. But now that firebreaks and roadways control fire, red and white pine are regaining dominance over closed-canopies of pioneer hardwoods, and on suitable soils beech and sugar maple have also gained in abundance. Both red and white pine are late successional species on sand, but on better, moister soils white pines speak of fire. Jack pines flourish on the sandiest and driest of soils, in sites most likely to burn. But paradoxically, and especially in the Upper Peninsula, jack pines also do well in sand-bottomed peatlands.

In northern Michigan, calcium-rich fens with moving, aerated groundwater eventually turn into white cedar swamps, but in poorly drained, oxygen-poor sites less rich in calcium, fens are transformed by a build-up of peat into acid Sphagnum bogs that get water and minerals from the atmosphere rather than from a mineral-rich groundwater. Cedar swamps and Sphagnum bogs are poorly developed under a more southerly climatic regime. In southern Michigan, with its summer heat and drought, productivity of biomass and accumulation of peat are much reduced. *Sphagnum*, scarcely favored by the environment, is less effective in acidifying fens and transforming them into bogs and black spruce muskegs. Our southern peatlands only rarely approach a bog level of development. What we call bogs are poor fens—poor in minerals but still under the influence of groundwater. Under relatively acid conditions, they take on a cover of tamarack and only rarely a scattering of black spruce, obviously dwarfed and disadvantaged so far from an optimal range in the Boreal Forest of Canada. More commonly, however, our fen mats develop a cover of red maple, tamarack, and poison sumac and become treed fens that speak of mineral nutrition from root level rather than from the air.

Habitat diversity is limited in southern Michigan, although a glacial topography provides a considerable up-down variation, and sandstone outcrops in a scattered few places, particularly in Eaton County along the Grand River. Lowered water tables associated with summer drought reduce the availability of moist sites or the wetness that surface ponding brings to boggy depressions. Most of the land has been taken over for agriculture, and woodlots are small and usually grazed. They are generally oak-hickory stands on sandy soils or hardwood swamps in lowlands, neither of much use to the farmer.

For what but picture galleries are the marble halls of these same hills?—galleries hung, month after month anew, with pictures ever fading into pictures ever fresh.

—Herman Melville

Southern Michigan cannot boast of the species richness that limy rocks would bring. It has, in fact, precious little rock, marble or otherwise, but beauty is there in plenty, changing with the calendar, and an up-hill, down-dale topography offers a gallery of habitats sufficient for many liverworts of charm, there for the finding in all seasons.

Uplands

Hot, dry summers conducive to fire have perpetuated prairie remnants from a post-glacial warm, dry period as parklike *oak-savannas*, scattered about, often on old beach ridges. In these "oak openings," well-spaced bur oaks and black oaks occupy a grassy dominance.

The upland *oak-hickory forest* is favored by a relatively warm climate over a long growing season and ready drainage. Summer drought and periodic burning favor oaks and hickories because they cope well with moisture stress and sprout vigorously after a burn. Dominant trees are black, white, and red oak, shagbark and pignut hickory, and black cherry. Characteristic associates are white ash, black walnut, hop hornbeam, and flowering dogwood. (American chestnut has virtually succumbed to the blight of disease.) The effective control of fire has allowed representatives from the moistness of beech-maple forests to move in and close the canopy.

The *beech-sugar maple forest* of better upland sites is favored by a long, warm growing season and soil moisture. Moisture has allowed these hardwoods to reproduce themselves and continue domination over long periods. The dominant beech and sugar maple are associated with basswood, white ash, tuliptree, red oak, black walnut, and bitternut hickory. Dense shading by a closed canopy allows little undergrowth. (This community differs from the northern hardwood forest of better upland sites in having a number of components not hardy farther north, tuliptree, for example. And also, hemlock and white pine are, and were also before the days of lumbering, poorly represented in comparison with northern Michigan.)

Wetlands

The *hardwood swamp* community occupies the post-glacial lake basin bordering Lakes Erie and Huron and also less extensive lowlands of glacial terrain, including a multitude of kettlehole depressions. These lowlands are cooled by cold air drainage, and a relatively constant water table makes dissolved nutrients continually available. Characteristic trees are black ash, red maple, yellow birch, pin oak, swamp white oak, and blue beech. (American elms, almost eradicated by disease, appear to be making something of a comeback in our swamplands.)

Bottomland hardwoods are subject to spring flooding and marked changes in the water table, as well as silting. The climate stays cool in the spring, owing to the slow warming of river water and the associated groundwater, as well as cold air drainage. Low springtime temperatures retard bud break and thereby protect trees from untimely frost. During the summer conditions become warmer and more humid than in the adjacent uplands. This extremely rich woodland includes many trees of the south—redbud, honey locust, Kentucky coffeetree, red mulberry, tuliptree, and sycamore among them. Other common species are black willow, cottonwood, silver maple, black walnut, butternut, and box elder. (American elm has, in the past half century, lost whatever dominance it once had because of disease.)

In wet depressions, on porous, gravelly till where a limy groundwater is aerated by motion, such as seepage, a *white cedar swamp* may develop. This kind of swamp has only a token occurrence in southern Michigan. More common are the *tamarack swamps* that develop where the groundwater is less calcareous and moderately acid. The well-spaced growth of tamarack allows the development of an understory of shrubs, with poison sumac invariably among them. *Sphagnum* is abundantly represented in the tamarack swamps.

Black spruce stands occur very sparsely in southern Michigan, in older parts of boglike poor fen mats and filled-in depressions, often encroached on and overtopped by red maple and a layer of shrubs. *Sphagnum* has a significant presence in such mats.

Open *rich fens*, or "marl bogs," dominated by sedgy plants, develop in calcareous wetlands, often in seepage areas. They have nothing to do with peatland successions leading to acid bogs. *Intermediate fens*, of mineral-rich, but less calcareous lowlands, often form sedgy mats at lake margins. They are commonly invaded by tamarack, poison sumac, and red maple and become *treed fens*, but under more acid conditions, they form *Sphagnum* lawns, or *poor fens*, and eventually support dense growths of heath shrubs, particularly leatherleaf and blueberries, and eventually some black spruce, at that stage approximating the acid bogs of the north.

The Habitat Preferences of Liverworts

Bryophytes generally grow in habitats conditioned by higher plants, in sheltered microhabitats where they escape from a competition with larger, rooted plants. They thrive on tops and sides of moist, shaded logs and stumps, on trunks and bases of trees, particularly in dark holes under buttress roots, on humus, often in seasonally flooded depressions and seepage around springs, and on banks of soil. Beech-maple woods, though moist, are too shady for most liverworts, and oak-hickory stands and oak savannas are too dry. However, banks of creeks and ravines, windfall mounds of soil, and glacial erratics provide habitat niches for some upland bryophytes. But it is in the wetlands that liverworts and mosses most abound, especially in hardwood swamps and river bottom woodlands. White cedar swamps, poorly developed as they are, feature some liverworts of characteristically northern distribution. Marly rich fens, sedge mats at the margins of small lakes, *Sphagnum*-dominated, bog-like fens and black spruce stands, though poor in species, also add a few elements of the north. Fields lying fallow after harvesting may, in wet years, give space to Riccias and hornworts. Rock outcrops, though exceedingly limited in Michigan's south, shelter some few bryophytes of interest.

The fact that our 75 species of liverworts are distributed into 44 genera means that they are taxonomically well separated, easily recognized, and taxonomically comfortable. Such a flora, with few species per genus, is good for beginners. It is somewhat less than half the size of the hepatic flora of the whole state. The species are, for the most part, common and widespread throughout the state and throughout eastern North America, and so the present book takes on a usefulness far beyond the limits of Michigan. The liverworts found south of the Tension Zone in Wisconsin and Minnesota are similar in number and in kind to those of southern Michigan, although both of those states have more rocky habitats and include "driftless areas" that escaped some of the rigors of glaciation. In comparable parts of Wisconsin there are 74 species of liverworts (Bowers & Freckmann, 1979), and in southern Minnesota there are 70 (Schuster, 1953). Certainly the liverworts occupying the flatness of Indiana, Illinois, and Ohio are little different from those of southern Michigan.

Steere's collecting activities, like my own, were mainly limited to a few counties centered around the University of Michigan and Michigan State University. He had access, however, to Irma Schnooberger's collections from many counties across the state from Port Austin to South Haven. His Michigan collections and Miss Schnooberger's too are stored at the University of Michigan. At Michigan State University are many fine collections made (especially) by John Engel, Norton Miller, Clifford Wetmore, and Ralph Common.

In the years that followed publication of Steere's liverwort book, Norton Miller recorded several interesting additions from Grand Ledge, and Michael Mayfield, Marie Cole, and Warren H. Wagner added a number of Riccias and hornworts from the southeast, many of them new to the state. Their success in finding was related to their hardiness. Not all of us are willing to go out at a time of year when deer hunting and frozen fingers detract from our pleasure.

Michael Penskar, William Brodowicz, and Dennis Albert have added a number of records of liverwort rarities, especially among the Riccias, hornworts, and Fossombronias. Joseph Jaworski, John Fody, Barbara Madsen, and Elwood Ehrle have taken a kindly interest in finding liverworts. Paul Redfearn, Janice Glime, Robert Belcher, and James Wells have made herbarium specimens available for study. Thomas Trana and Irene Crum have helped me to exploit the resources in the liverwort herbarium at Michigan State

University. Norton Miller solved a few problems in identification, and Karen Sue Renzaglia helped me to a better understanding of hornwort morphology.

William Buck, Bill McKnight, Dennis Hall, and Joseph Jaworski made useful criticisms on the manuscript, as did Warren H. Wagner, Elwood B. Ehrle, and John Fody. Florence Wagner and David Bay helped with photography.

Many of the illustrations are my own. Marie Wohadlo's stylish habit sketches were made possible by bits of money from the Department of Biology, the University Herbarium, the Faculty Fund of the College of Literature, Science, and the Arts, and research funds from the Office of the Vice-President for Research, all of the University of Michigan. Wendy Zomlefer prepared the title page design.

Honorable mention goes to William R. Anderson, Director of the University Herbarium, for being a good friend and clever fund finder. I am greatly indebted to William J. Gilbert and Warren H. Wagner for a grant-in-aid of publication from the Hanes Fund. The costs of publication were supplemented by the Gillette Fund administered by the University of Michigan Herbarium.

Profuse and sincere thanks go to Christiane Anderson for her pleasant patience and skill in editing the manuscript.

Much of the work on the book was done at home. My wife, Irene, was pleasant on the many evenings and weekends when my thoughts were weighted toward an enjoyment of liverworts. In her long association with me, she sees the sense of Robert Frost's doggerel:

> It takes all sorts of in and outdoor schooling
> To get adapted to my kind of fooling.

Taxonomy of Liverworts and Hornworts

Though easily parceled out as mosses and liverworts, the bryophytes, of the division Bryophyta, group themselves more naturally into five classes: true mosses (Bryopsida), granite mosses (Andreaeopsida), peat mosses (Sphagnopsida), liverworts (Hepaticopsida), and hornworts (Anthocerotopsida). In spite of their small size and relative simplicity, bryophytes are greatly diversified, and the component classes are well separated and only distantly related to one another. To a bryologist, liverworts seem vastly unlike hornworts and certainly unlike any of the mosses. Yet resemblances in life cycles and reproductive structures suggest a similar, though not necessarily identical ancestry. The ultimate origins of bryophytes, among aquatic green algae, and the earliest invaders of land far outdate the fossil record. The oldest known representatives of each class of bryophytes seem elaborately differentiated and quite modern. They are perfectly distinct at the class level, and no intermediates link either fossil or modern members of those classes.

Bryophytes are small and, relative to other land plants, simple. They lack roots and have no vascular system at all or only a suggestion of conductive tissues. Although most of them are perennial, they have no secondary growth, and (except in the sporophytes of hornworts) they grow by means of single cell initials rather than meristematic tissues. They have no structural provision for winter or dry season dormancy, and the embryo grows directly into a sporophyte (rather than remaining dormant in a seedlike protective structure). Like all land plants, bryophytes have sexual and asexual generations in alternation. The haploid, gamete-bearing generation that begins with spore germination is the more dominant and complex phase of the life cycle. It may be simple and thallose (as a forked ribbon or a rounded rosette) or more complex and leafy. It is anchored to the substrate by hairlike rhizoids. The sex organs, antheridia and archegonia, may occur on the same or different plants.

Flask-shaped archegonia contain a single egg. Spherical, ellipsoidal, or banana-shaped, stalked antheridia produce countless biflagellated sperm cells. Both archegonia and antheridia are multicellular structures protected by a jacket layer of sterile cells. The diploid, spore-bearing generation results from the fertilization of an egg. The sporophyte grows attached to the gametophyte, as a dependent, even as a parasite. Relatively small and simple, it completes most or all of its development inside the enlarged base of the archegonium. Radially organized and unbranched, the sporophyte normally consists of an absorptive foot enclosed in gametophytic tissue, a stalklike seta, and a capsule in which reduction divisions take place and large numbers of spores are formed. On germination, a spore forms a juvenile stage, or protonema (greatly reduced or essentially lacking in liverworts and hornworts), and eventually into a thallose or leafy gametophyte.

Hornworts are commonly regarded as a special kind of liverwort. Both hornworts and liverworts have flat gametophytes with 1-celled rhizoids, scarcely any protonema, and erect capsules. There similarities end, and differences crowd into view:

Liverworts (p. 17) are thallose or leafy and have sex organs of superficial origin. Chloroplasts are many and small. The sporophytes, short-lived and parasitic on the gametophyte, lack a meristematic tissue. The capsules, nearly always elevated on a seta, have no columella or stomata, and they dehisce, most commonly, by splitting into four valves. The spores are generally mingled with slender, spiral-thickened, hygroscopic elaters. The spores are simultaneously matured and dispersed.

Hornworts (p. 105) have gametophytes in the form of a flat rosette and produce sex organs internally. The chloroplasts, generally one per cell, are large and platelike with a central pyrenoid group that accumulates starch (as in green algae). Lacking a seta, the sporophytes typically consist of a massive foot and a long, green capsule that splits into two valves at the upper end while continuing growth from the base. The capsule wall consists of an epidermal layer of long cells and, generally, stomata embraced by two kidney-shaped guard cells. The cells of the inner, solid tissue and the guard cells, at the surface, generally contain two chloroplasts. The spores, surrounding a slender columella, mature progressively toward the base of the capsule and are therefore gradually dispersed. They are mingled with pseudelaters consisting of one to several cells and usually lacking spiral thickenings.

A basic distinction is that liverwort spores derive from the endothecium [fig. 13a–c], whereas those of hornworts form in the amphithecium [see fig. 102e–g]. (The first divisions of a young capsule form a quadrant, or group of four cells, each of which divides to set off an outer amphithecium and an inner endothecium.)

It is not my intention to present here a coherent essay on morphology but rather to insert bits of information, together with drawings, in a taxonomic context. Additional information on structure and reproduction is provided by Schuster (1949, 1953, 1966–81, 1983–84), Müller (1905–16, 1951–58), Macvicar (1926), Watson (1971, 1981), Renzaglia (1978), Crandall-Stotler (1984), Flowers (1961), Casares-Gil (1919) and Schofield (1985).

> Thou sayest an undisputed thing in such a solemn way.
> —Oliver Wendell Holmes

Schuster's elegant books on the liverworts of New York and Minnesota are very useful for identifications anywhere in temperate eastern North America, including Michigan. His 6-volume work on the liverworts and hornworts of North America (two volumes not yet published) is too detailed for casual use, but his illustrations will please the veriest tyro as well as sophisticates suffering from a heavy-lidded *Weltschmerz*. As a supplementary guide, the treatment of liverworts and mosses of eastern Canada by Ireland & Bellolio-Trucco (1987) deserves commendation. Conard & Redfearn's *How to Know the Mosses and Liverworts* is similarly helpful anywhere on the continent. (The trouble with both of the latter is that they present, admirably, some, but not all species.) Surprisingly enough, Macvicar's *Student's Handbook of British Hepatics*, old as it is, can be used to advantage because of a considerable overlap in species as well as straightforward descriptions and pertinent drawings. (As many as 80% of the liverwort flora of southern Michigan appear in Macvicar's *Handbook*.) Watson's *British Mosses and Liverworts* and, especially, Smith's *The Liverworts of Britain and Ireland* are also good to have in hand. I enjoy and profit by an occasional re-reading of Evans' "Notes on North American Hepaticae" (1910–23) and "Notes on New England Hepaticae" (1902–23). In making use of a diversity of literature, the checklist of North American liverworts and hornworts (Stotler & Crandall-Stotler, 1977) will bring names, old and new, into concordance.

Class Hepaticopsida (Liverworts)

The gamete-bearing plants, or gametophytes, are dorsiventrally organized as undifferentiated thalli or leafy shoots anchored by unicellular rhizoids. Thalli and thallus lobes are variously thickened, but leaves are one cell thick and lack a midrib. The cells contain numerous small chloroplasts and usually oil bodies too. The sex organs, antheridia and archegonia, are superficial in origin (but in some cases come to be enclosed in air chambers). The spore-bearing plants, or sporophytes, reduced in size and complexity, are attached to the gametophyte and parasitic on it. The sporophytes are enclosed until maturity in the enlarged base of the archegonium, or calyptra [figs. 2, 12c], which is, in most species, finally ruptured as the sporophyte increases in size. The seta, greatly reduced or even lacking in some thalloid genera, commonly elongates at maturity as a delicate, short-lived stalk. The capsule has a solid wall, without stomata, and lacks a columella. The spores, derived from the endothecium and occupying the entire capsule interior to the wall, are generally mingled with long, slender, single-celled elaters with spiral thickenings.

The female sex organ, or archegonium [figs. 12d, 33a, 38a], consists of a swollen base, or venter, containing a single egg and a long, narrow neck that at maturity encloses a canal through which the sperm swims to the egg. The egg is never released. The male organ, or antheridium, is a stalked, generally spherical structure with a jacket of cells enclosing a vast number of sperm cells [figs. 3, 11, 38b] that are explosively discharged in the presence of water. A substance released with the sperms reduces surface tension and permits a rapid spread over the surface of a water film. Each sperm has an elongate, spiral body with two flagella at one end. Once in the vicinity of a mature archegonium (with an open canal), the sperm ceases its aimless gyrations and swims toward the egg. The fertilized egg, or zygote, apparently releases a substance that stimulates the venter to grow into a calyptra and surrounding tissues to form structures that sheath and protect the developing sporophyte.

Antheridial development [figs. 9, 10] begins with the transverse division of a superficial cell to set off a basal cell giving rise to a stalk from an upper cell that becomes the sperm-producing body. Transverse divisions of the upper cell result in a filament of four to six cells. In the Marchantiales each cell of the filament undergoes vertical divisions at right angles to form a quadrant of cells (as seen in cross-section but actually several in a vertical series). The cells of the quadrant then divide tangentially to delimit a group of four primary androgonial cells surrounded by four jacket initials. In the Metzgeriales and Jungermanniales no quadrant is formed and only two androgonial initials are enclosed by four jacket initials.

Archegonial development [fig. 12a–b] also involves superficial cells that divide to form a vertical file of two or three cells. The uppermost cell undergoes three excentric vertical divisions enclosing a central cell exposed at its upper end until a transverse wall delimits a cover cell. In the Marchantiales, each of the three jacket initials divides vertically so that the archegonial jacket consists of six cell rows. In the Metzgeriales and Jungermanniales, by contrast, only two of the jacket initials divide, and the jacket at maturity thus consists of five rows of cells.

The embryo grows first into a ball of eight cells or a 3- to 4-celled filament. After a short period of generalized cell divisions, foot, seta, and capsule differentiate [fig. 33c–d]. Early on, in the capsule area, an outer amphithecium is differentiated from an inner endothecium [fig. 13a–c]. The amphithecium gives rise to the capsule wall, and the endothecium becomes sporogenous tissue [fig. 13b–d]. It is in the capsule that reduction divisions take place and spores are formed [fig. 13e].

The spores are generally interspersed with slender, spirally thickened elaters [fig. 25d]. The elaters are filled with water that comes under evaporative tension on exposure to dry air. Tension on the spiral thickenings causes the elaters to quiver and spores to be dispersed. More violent motions may cause both spores and elaters to be hurled into the air. In most liverworts the elaters lie free within the capsule, though frequently oriented longitudinally or obliquely. In the leafy liverwort Lejeuneaceae and Frullaniaceae, the elaters are attached to the capsule wall at both ends but become free at the lower end when the capsule dehisces. In various of the Metzgeriales, some of the elaters are attached to a tufted mass of stout elaters [figs. 38c, 39e, 40f], either at the base or the apex of the capsules.

Each cell of the sporogenous tissue divides into two daughter cells, one of which elongates as a diploid elater. The other usually undergoes a series of transverse mitotic divisions to form a file of spore mother cells before undergoing meiosis [fig. 13e]. If it undergoes meiosis directly [fig. 13d], four haploid spores are formed for each elater, in a ratio of 4:1, but if it divides mitotically one or more times before reduction takes place, many more (and smaller) spores are formed in a spore:elater ratio of 8, 16, or 32:1. In *Marchantia polymorpha* the ratio may be, on occasion, as much as 128:1. In *Schistochila*, a leafy liverwort, as many as 5,000,000 spores per capsule and a spore:elater ratio of 240–360:1 have been recorded.

Fertilization results in a single offspring sporophyte which, however, produces a profligate number of spores and genetically differentiated gametophytes. But sexual reproduction is, in many species, far less important than the vegetative duplication of populations by minute gemmae borne on leaf tips, in stem tip clusters, or in cuplike receptacles, by deciduous leaves and branches, and rarely by casual fragmentation followed by regeneration.

In most cases, apparently, a germinating spore produces only one mature gametophyte. The protonema and initiating gametophore, together making up the sporeling stage, exist in many forms, only a few of which are illustrated here [fig. 4]. Patterns of sporeling development correlate with certain genera and families and thus appear to have some phylogenetic significance. The protonema (so important in the formation of extensive moss clones) may be no more than a short filament or cell mass that persists only until an apical cell is formed. However, in some genera, among them the leafy *Lophocolea* and *Chiloscyphus* [fig. 4g] and the thallose *Bucegia* [fig. 4h], a potential for cloning is provided by a branched, filamentous protonema. (Especially among leafy liverworts, clones commonly result from the dying away of older portions of plants.)

> I will discuss this chapter with the following observation: that it becomes an author to divide a book, as it does a butcher to joint his meat, for such assistance is of great help to both the reader and the carver.
>
> —Henry Fielding

The liverworts consist of six "joints," or orders: Calobryales (often called the Haplomitriales), Metzgeriales, Jungermanniales, Sphaero-carpales, Monocleales, and Marchantiales. (The Takakiales, discovered only 40 years ago, are a highly distinctive group better referred to the mosses.) In Michigan, we have only the Metzgeriales, Jungermanniales, and Marchantiales. The following text is "jointed," as Fielding would say, according to those orders, and the hornworts are appended at the end. The families are used in this work only for grouping and arranging genera. (They are not described here because they are based on generalizations difficult to understand on a local flora basis.) The text is not phylogenetic in organization. The simplest and most reduced liverworts are placed in the beginning even though their simplicity speaks of a derived rather than a primitive stage of development.

1. Plants leafy.   Jungermanniales, p. 51
1. Plants thallose (except that *Fossombronia* in the Metzgeriales has a leafy construction).
   2. Thallus dichotomously forked, consisting of a loose upper tissue and generally having dorsal pores and ventral scales.   Marchantiales, p. 20
   2. Thallus not dichotomously forked, consisting of solid tissue and lacking pores and scales.   Metzgeriales, p. 39

ORDER MARCHANTIALES

The gametophytes grow as dichotomously forked ribbons or rounded rosettes. The rhizoids are of 2 kinds, smooth and internally pegged, and rows of unistratose ventral scales are usually present. (Rhizoids and scales are lacking in aquatic plants.) The thallus shows internal differentiation into a solid, colorless to purplish lower tissue and a loose, green upper tissue made spongy by air chambers or narrow channels among crowded, erect filaments. Pores on the upper surface may be simple or well marked by a stomatal structure of differentiated cells in 1 to several layers. The sex organs, though superficial in origin, may come to be enclosed in cavities of the thallus and scattered at the surface or aggregated on male or female receptacles that are sometimes elevated on stalklike branches. In the latter case, the antheridia are enclosed in cavities, but the archegonia are exposed and pendent from tissues situated in the sinus between rays of the female receptacle. The archegonia have 6 rows of neck cells. The sporophytes, small and reduced in complexity, may consist of a mere globose capsule producing only spores or an ellipsoid capsule provided with a short, massive seta and a large foot and producing elaters as well as spores. The capsule wall consists of a single layer of cells, and dehiscence occurs irregularly, though sometimes by a lidlike operculum or longitudinal divisions, or valves.

Riccia and *Ricciocarpos* produce their sporophytes in air chambers [fig. 20b–c] or narrow, vertical air channels [fig. 18c]. The sporophytes consist of nothing more than a globose capsule with a wall of one cell layer. The delicate cells making up the wall are one layer thick and lack thickenings. All cells internal to the wall become spores [fig. 2].

Other members of the order, such as *Conocephalum* or *Marchantia*, are more complex: Their gametophytes may have simple pores or compound stomatal structures made up of differentiated cells, and their sex organs are grouped on receptacles that are often elevated on stalks [figs. 32, 34b–c]. The stalk is a modified branch as evidenced by rhizoid furrows often developed on the morphologically ventral side of such structures. Archegonia are initiated on the receptacles at marginal growing points that come to be decurved because of unequal growth of the upper and lower tissues. As a result, the archegonia have a displaced, ventral position, and the youngest archegonia are located near the stalk of the receptacle and older ones in a progression toward the outside [fig. 33a]. The sporophyte consists of a massive foot, a short, thick seta, and an ellipsoid capsule containing both spores and spiral-marked elaters. The cells of the capsule wall are strengthened by complete or incomplete annular thickenings. The sporophyte develops within a calyptra derived from the expanded base of the archegonium, or venter [figs. 2, 33a], as well as an involucre formed by an upgrowth of thallus tissue [figs. 25e, 33a]. Some genera have, inside the involucre, a pseudoperianth surrounding one or more archegonia and sporophytes [fig. 25e]. Among our genera, the pseudoperianth of *Asterella* is conspicuously large [fig. 29b]. It is less conspicuous in *Preissia* and *Marchantia* and lacking in *Reboulia*, *Conocephalum*, and *Lunularia*.

The spore mother cells remain unlobed, in contrast to those of the Metzgeriales and Jungermanniales where they become deeply 4-lobed before meiosis. The spores typically have a convex outer face and three flat inner faces separated by triradiate ridges.

Gametophytic growth never results from the activities of a tetrahedral apical cell with three cutting faces, as in most other liverworts, and the thallus is accordingly not triradial in organization. The apical cell has instead four cutting faces, two lateral, one dorsal, and one ventral. The Marchantiales and some of the thallose Metzgeriales appear to branch by equal forking, or dichotomy. A true dichotomy, as in the hornworts [fig. 6b–c], results when an apical cell divides into two cell initials that come to be separated by a group of inactive cells. In the Metzgeriales forked branching appears to be pseudodichotomous, the apical cell giving rise to derivatives among which a second apical cell soon differentiates [fig. 6d]. Branching in the Marchantiales may seem to result from paired apical cells, but whether it is truly dichotomous or not is unlikely.

The growing point, in notches of the thallus, is protected from drying by mucilage (or slime) hairs [fig. 5]. Sex organs may also be protected by slime-producing mucilage hairs [fig. 9d].

The oil bodies of the Marchantiales are fairly uniform and therefore of little taxonomic interest. They are solitary bodies nearly filling scattered oil cells [figs. 31b, 35b] that are somewhat smaller than the other cells.

1. Plants aquatic, though commonly stranded during part of the year.
  2. Plants floating just below the surface, consisting of repeatedly forked, linear segments and lacking ventral scales and rhizoïds.   1. *Riccia*, in part
  2. Plants floating on the surface, consisting of a heart-shaped or fanlike thallus with scales and rhizoids on the underside.   2. *Ricciocarpos*
1. Plants terrestrial (see also *Riccia* and *Ricciocarpos* for stranded forms).
  3. Thalli small and rosettelike or, less commonly, consisting of slender, forked ribbons.   1. *Riccia*, in part
  3. Thalli larger, coarser, branched ribbons, generally dichotomously forked.
    4. Gemmae cups usually present.
      5. Plants of greenhouses, short, broad, and not much branched; gemmae cups bordered on 1 side by an entire, crescent-shaped rim.   3. *Lunularia*
      5. Plants of greenhouses and also wet, outdoor habitats, elongate, conspicuously forked; gemmae cups bordered by a fringed, circular rim.   8. *Marchantia*

4. Gemmae cups none.
>   6. Plants large and broad, without red pigmentation; upper surface conspicuously marked by a coarse pattern of air chambers, each with a large, white stomatal structure.   4. *Conocephalum*
>   6. Plants of moderate size, reddish (at least at the margins and on the undersurface); air chambers and stomatal structures less conspicuous.
>   > 7. Epidermal cells with thickened corners; stalk of the female receptacle scaly-fringed at top and bottom.   5. *Reboulia*
>   > 7. Epidermal cells without thickened corners; stalk of female receptacle not fringed at top and bottom.
>   >   > 8. Thallus margins strongly incurved when dry; upper surface smooth and pores inconspicuous; female receptacles conic; pseudoperianths large, much divided to form a basketlike fringe, usually reddish above, yellow below.   6. *Asterella*
>   >   > 8. Thallus margins not strongly incurved; upper surface minutely areolate and stomatal structures small, though distinct; female receptacles square, 4-ridged at top; pseudoperianths not much divided.   7. *Preissia*

RICCIACEAE

1. **Riccia** L.

The Riccias grow as slender, forked ribbons or small, round rosettes. They are annual plants; some are aquatic but most of them grow on disturbed soils. Scales and rhizoids occur on the underside of terrestrial plants; the scales are in 2 rows near the growing points. The thallus consists of a solid lower tissue and a spongy upper layer of air chambers separated by plates of cells or narrow vertical spaces among columns of cells. Pores on the upper surface, if present, are surrounded by undifferentiated cells. Oil bodies are lacking. The sex organs, scattered along the thallus middle, come to be deeply immersed. The sporophyte consists of a globose capsule lying inside the enlarged venter of the archegonium. The capsule wall of 1 cell layer deteriorates before spore maturation, and the spores accordingly lie free in the calyptral cavity. Eventually the thallus rots away to expose dark, globular spore masses. The spores are large and tetrahedral, and elaters are lacking.

The generic name gives honor to Pietro Francesco Ricci, a Florentine senator. In his pre-Linnaean *Nova Genera Plantarum*, Micheli acknowledged Ricci's support on a plate illustrating numerous species that he assigned to the genus *Riccia*.

Riccia bifurca, R. beyrichiana, and *R. hirta* (subgenus *Riccia*) have relatively solid thalli with narrow air channels among columns of green cells [fig. 18c]. (Each column is commonly topped by a pear-shaped, colorless cell that becomes collapsed and flat.) By contrast, *Riccia fluitans, R. rhenana, R. canaliculata, R. sullivantii*, and *R. cavernosa* (subgenus *Ricciella*) are made spongy by sizable air chambers [fig. 15b, e], and all of the latter group except *Riccia cavernosa* have ribbonlike thalli producing spores in ventral protrusions [fig. 17b]. Both kinds of air spaces, whether cavities or narrow channels, result from schizogeny, the splitting apart of adjacent cells of the thallus tissue. (The several-layered air chambers of *Ricciocarpos* are likewise formed schizogenously, but early in their development they become subdivided by cellular partitions.) The sex organs are to be seen individually inside such spaces [fig. 19a–b] scattered along the median grooves. Actually initiated at the dorsal surface close behind the growing point, they are quickly enveloped by the upgrowth of surrounding tissues. The tip of the archegonium may project slightly above the surface [fig. 2]. The antheridia are situated below the surface, but the site of the antheridial chamber is, in many species, marked by a slender, projecting ostiole.

The capsule wall is delicate and lacks thickenings. It is used up in nourishing spore mother cells and developing spores, in both *Riccia* and *Ricciocarpos*, and at maturity, the spores lie in a globose mass inside the calyptra [fig. 2].

Riccias have been found in southern Michigan only sparsely and almost exclusively in the fall of the year, but they can also be expected in early spring before competition from larger plants (especially weeds) sets in and even in the summer in places unseasonably wet, such as depressions in stubble fields and muddy margins of streams and lakes exposed by lowered water levels.

1. Thallus without conspicuous air chambers, not areolate at the surface.
    2. Thallus green or, in the sun, red, with fleshy, grooved segments; marginal cilia none; spores 80–100 μm wide.   4. *R. bifurca*

    2. Thallus gray-green, not fleshy or conspicuously grooved; margins ±
       ciliate, especially at tips of segments; spores 90–130 μm wide.
       3. Cilia usually numerous at apical and often lateral margins, stout, 75–
          300 μm long; primary segments 1.5–2.5 mm wide.   6. *R. beyrichiana*
       3. Cilia few, restricted to tips of young branches, less than 100 μm long;
          primary segments about 1 mm wide.   5. *R. hirta*
1. Thallus with well-developed air chambers, areolate.
    4. Thallus slenderly ribbonlike, the segments less than 1 mm wide.
       5. Plants sterile, floating or stranded on mud, repeatedly branched, the
          branches not thick or grooved.   1. *R. fluitans*
       5. Plants often fruiting, terrestrial, little branched, the branches thick
          and fleshy, grooved toward the tips.   [*R. canaliculata*]
    4. Thallus rosettelike, the segments generally more than 1 mm wide.
       6. Thallus pale, frosted-green, in round rosettes with broad lobes, con-
          spicuously pitted-lacunose; spores 67–84 (110) μm wide.
                                           3. *R. cavernosa*
       6. Thallus dark-green (often with red tinges), in incomplete rosettes
          with narrow, forked segments, sometimes lacunose in older parts;
          spores 44–78 μm wide.   2. *R. sullivantii*

1. *Riccia fluitans* L. [fig. 15a–c] is a slender, dark, though translucent, green aquatic consisting of a thin, repeatedly forked ribbon
with linear segments less than 0.7 mm broad. There is no dorsal
groove, and rhizoids and rudimentary scales are seen only in
stranded forms. The plants are dioicous. Capsules (not known from
North America) are produced in a ventral position. The spores measure 50–80 μm in diameter.

The specific epithet means floating. The plants float just below
the surface of quiet pools, less often in slowly running water. In late
summer, they often become stranded and take on a firmer, thicker
appearance. They have been found in Allegan, Barry, Branch, Clinton, Eaton, Genesee, Gratiot, Hillsdale, Ingham, Jackson, Kalamazoo, Kent, Livingston, Macomb, Monroe, Oakland, Tuscola, and
Washtenaw Counties.

A specimen from Eberwhite Woods in Ann Arbor (only a block
from my home) was determined by Schuster as *Riccia rhenana* K.
Müll. with a notation that the chromosome number is n=16+2m
(twice that of *R. fluitans*). (Hillsdale County plants of similar size
and appearance have also been seen.) *Riccia rhenana* has segments
somewhat more than 1 mm wide, sometimes as much as 2 mm, and

cells larger, those at thallus margins being about 25×40–50 μm, in contrast to R. *fluitans*, with segments about 0.7 mm wide and marginal cells only about 12×30 μm. This species is not easily identified even on careful comparisons, and so it suits our convenience to consider it a mere diploid of R. *fluitans* rather than a separate species. However, McGregor (1952) kept Kansas plants of R. *fluitans* and R. *rhenana* in culture for three years during which time they changed from land to water forms and back again several times. The water forms resembled each other more than the terrestrial ones, but the species could always be distinguished. Both remained sterile in both aquatic and stranded forms.

It is reported that plants of *Riccia fluitans* and R. *rhenana* have been dispersed by snapping turtles and ducks. A common habitat for terrestrial Riccias is wet soil trodden by cattle or ruts made by farm machinery at corners of fields. It is said that in Hungary spores of the terrestrial R. *frostii* Aust. [fig. 16a] are carried on the muddy bills and feet of geese, and the species is accordingly found especially along goose paths. That species of wide distribution, to be expected in our range, has been found in northern Michigan, in Delta County, on riverside mud exposed by lowered water levels. (The species forms a rosette of narrow segments that are flat except for a poorly marked median groove. Thalli that have not yet ceased to grow at the tips already have older tissues ruptured and lacunose-pitted. The spores are relatively small, about 45–60 μm in longer diameter, and irregularly ridged and anastomosed but not regularly reticulated.)

Riccia canaliculata Hoffm. [fig. 15d–g], a terrestrial species formerly considered to be the land form of R. *fluitans*, has shorter, broader segments, about 1 mm wide and somewhat grooved. The plants are monoicous. Spores, commonly produced in a ventral position, measure 63–80 μm (and supposedly as much as 100 μm). The species, apparently rare in North America, has been reported from a fallow corn field in Wayne County (Mayfield et al., 1983). The identification could not be confirmed.

2. *Riccia sullivantii* Aust. [fig. 17a–c] forms rounded rosettes or overlapping dichotomies that are slender, though firm and fleshy, with branches up to 1.5 mm wide, blunt or truncate at the tips, and

somewhat grooved. The dark-green thalli (larger but much resembling the land form of *R. fluitans*) may be reddish at the margins and in lower tissues, and they sometimes become ± lacunose-pitted with age. The plants are monoicous, and the spores, produced in the autumn in ventral swellings, are 44–78 μm wide and reticulate on the outer face.

The species was named for the distinguished bryologist William Starling Sullivant of Columbus, Ohio.

The species has been reported from Allegan, Gratiot, Monroe, and Washtenaw Counties, on wet soil in fallow fields and on organically enriched sand in swales and borrow pits, at lake margins, and on disturbed soil of 2-track roads.

3. *Riccia cavernosa* Hoffm. (*R. crystallina* of American authors) [fig. 17d] forms pale, frosted-green rosettes with sparse dichotomies and short, broad segments (1.5–2.5 mm wide) rounded or truncate at the tips. The upper epidermis regularly deteriorates to expose a honey-combed interior. The plants are monoicous, and the capsules are included in the thallus. The spores are 60–120 μm wide and incompletely reticulate on the outer face, even more so on the inner. Spores in autumn.

The specific epithet refers to a thallus riddled with cavities.

Plants have been encountered in Monroe County, on wet soil in fields harvested for corn and wheat (Mayfield et al., 1983) and also on wet, disturbed soil of a 2-track road in Huron County.

4. *Riccia bifurca* Hoffm. (*R. arvensis* Aust.) [fig. 18h–k] forms small rosettes or irregular mats that are dull and grayish green or, in the sun, reddish, at least at the margins. The segments, about 1.5 mm wide, are thick and fleshy on both sides of a fairly broad, flat-bottomed groove that is often closed in at the blunt lobe tips. The youngest segments are short, widely divergent, and lacking in marginal cilia. The plants are monoicous. The black spores, measuring 65–110 μm (averaging about 85 μm), are reticulate on both inner and outer faces and winged at the junction of convex and plane surfaces. Spores in late autumn.

The specific name refers to forked thallus lobes.

Plants have been found Monroe and Wayne Counties, on the wet soil of old corn fields and winter wheat fields.

5. *Riccia hirta* (Aust.) Underw. [figs. 16b,18e–g] forms rounded rosettes or intricate mats that are a dull gray-green but with age become dark purplish red. The thalli are usually 2–3 times dichotomous, and the segments, rounded at the tips, are only about 1 mm wide. The margins tend to curve upward, and the segments are accordingly rather concave but not noticeably grooved or only narrowly so toward the lobe tips. The lobe tips show few to rather numerous (and often deciduous) cilia measuring about 30–120 μm long. The plants are monoicous, and the spores are brown, 88–110 μm or more in width, reticulate on both inner and outer faces (but more regularly so on the outer), and winged at the junction of the outer convex face and the 3 inner plane faces. Spores in autumn.

The plants have been found in Monroe, Washtenaw, and Wayne Counties, on wet soil in fields harvested for corn, soybeans, and wheat.

As contrasted with *R. beyrichiana*, this species has more open branching and narrower segments. It takes its name from the ciliate margins of thallus lobes. The cilia are shorter and finer than those of *R. beyrichiana* and often few or deciduous. (Species with marginal cilia lose them under conditions of unusual humidity, in culture, for example. Conversely, plants of relatively dry conditions have a good development of cilia.)

6. *Riccia beyrichiana* Lehm. [figs. 16c, 18a–d] forms rather coarsely dichotomous-branched, round rosettes that are reddish to blackish purple in dry or exposed sites. The chief segments are 1.5–2.5 mm wide, the ultimate ones about 1.5 mm in width. The segments are not very concave or grooved but have near their tips rather broad, shallow depressions that are abruptly closed in at the apex. Young portions of segments are fringed by stout cilia varying in length from about 75 to 300 μm. The plants are monoicous, and the brown spores are about 88–130 μm in width, winged at the junction of inner and outer faces, and reticulate on their outer faces. Spores in spring to late fall.

This species was first found in Georgia by a German collector, Heinrich Karl Beyrich.

The plants have been found in Monroe, Washtenaw, and Wayne Counties, on the wet soil of fields where corn and soybeans have been harvested, also among winter wheat and in ditches at the side of roads and at the bottom of railroad grades.

In contrast to *R. hirta*, the thalli have more crowded segments with a broad depression near their tips and more numerous, stouter cilia.

2. **Ricciocarpos** Corda

These rather small, silver-green, heart- or fan-shaped plants float on the surface of ponds and show around them a fringe of long, purplish black scales produced on the undersurface. Rhizoids are few or lacking except in the terrestrial form, in which scales are much reduced. Each lobe has a dorsal groove at its middle. The thallus consists almost entirely of a green, spongy tissue made up of several layers of air chambers. The pores on the upper surface are surrounded by not or slightly differentiated cells. The plants are monoicous, but antheridia form earlier than archegonia. The sex organs [fig. 20a–b] are initiated superficially but come to be enclosed in chambers scattered the length of the dorsal grooves. The sporophyte is no more than a ball of spores inside a capsule wall of 1 layer of cells. (There are no elaters.) Disintegration of the wall leaves the spores lying free in the enlarged venter of the archegonium, or calyptra [fig. 20c–d]. The tetrahedral spores are about 50 μm wide, blackish, and spiny.

The genus takes its name from *Riccia*-like capsules, or "fruits," visible as blackish masses beneath the median grooves.

Ricciocarpos natans (L.) Corda [figs. 19-21; *frontispiece*] floats on the surface of water, not just below it as *Riccia fluitans* does. The thallus consists almost entirely of large air chambers that give it such buoyancy. The fan-shaped plants grow in shallow swamp pools, often in the backwaters of streams, but as water levels fall during the dryness of summer, the plants "root" on the mud and take on the form of a 1-sided rosette [fig. 21f] with firm, narrow segments (about 2–3 mm wide) and numerous rhizoids but few ventral scales. (They can be taken for *Riccia*, but there is a narrow, though distinct dorsal groove, and the epidermis shows a fair degree of sectoring). In springtime, when pools are full, the floating form produces sex organs and matures spores in May and June. Black spore masses are easily seen through the thallus tissue.

Collections have been made in Barry, Branch, Genesee, Gladwin, Gratiot, Ingham, Isabella, Jackson, Kalamazoo, Kent, Lenawee, Macomb, Monroe, Sanilac, St. Clair, Tuscola, Washtenaw, and Wayne Counties.

The plants, named in reference to floating, or swimming, form a crowded surface mat in quiet pools, usually in association with duckweeds. Their rhizoids and scales often wiggle with an abundance of invertebrates.

Plants stranded on mud in the late summer become submerged on the advent of fall rains and overwinter in submergence. All but the growing points may die as a result of drying in late summer or freezing and thawing in winter, but the growing tips often remain viable. In spring the tips of thalli become detached, float to the surface, and grow into fan-shaped, fruiting individuals. The landed forms have longer, narrower, forked segments in irregular, 1-sided rosettes. They are likely to be taken for a *Riccia*, but there is a narrow, though distinct dorsal groove. The epidermis shows some sectoring (in correspondence with the underlying air chambers), as well as inconspicuous pores and oil cells. Rhizoids are numerous, but ventral scales are much reduced.

LUNULARIACEAE

3. **Lunularia** Adans.

The thallus is a short, broad ribbon of a glossy, dark-green except for hyaline margins. Gemmae are borne in half-rimmed depressions with entire margins. The upper surface is rather distinctly areolate, and the stomatal structures, though small, are easily discerned. The pores are surrounded by several rings of hyaline cells in a single layer. The 1-layered air chambers are bottomed by short, green filaments. The solid ventral tissue includes numerous oil cells that are smaller than other cells and nearly filled with a single oil body; the ventral cells often have faintly punctate-striate walls. The green lower surface bears scales at both sides of the midrib; they are delicate, hyaline, and broadly lunate, each with a rounded appendage. The plants are dioicous. The male receptacles are sessile discs. The female receptacles, elevated on a subhyaline, ± hairy stalk lacking a rhizoid furrow, seem to be lobed nearly to the base, but actually they consist essentially of 4 tubular involucres horizontally spreading in a cruciate arrangement. Each of the involucres encloses 1 or sometimes 2–3 sporophytes. There is no pseudoperianth between involucre and calyptra. The seta is somewhat elongated at maturity, and the capsules split nearly to the base into 4 valves.

The generic name, meaning little moon, refers to the crescent-shaped rim of the gemmae cups.

Lunularia cruciata (L.) Dum. [figs. 22, 23] is a greenhouse weed fruiting in North America only in California, where it escapes to the out of doors. (It also completes its sexual cycle in the Mediterranean region, western continental Europe, the south of England, and New Zealand.) Plants have been seen in the conservatory of the University of Michigan Botanical Garden, in Washtenaw County, and in a Jackson County flower shop.

The specific epithet bears reference to the 4-rayed, crosslike female receptacle.

CONOCEPHALACEAE

4. **Conocephalum** Wigg.

This large, ribbonlike liverwort is recognized by coarse sectoring and bulging, white pore structures of its upper surface. The 1-layered air chambers are floored by short, green filaments ending in a long-pointed, hyaline cell. The pores are surrounded by a mound of 5–7 concentric series of hyaline cells in 1 layer. Cells of the compact lower tissue have walls with fine, vertical striations. Tubes running lengthwise of the thickened thallus middle show in section as large, round holes. The undersurface is pale-green and has rather distant, delicate scales in 1 row on each side of the midrib, each with a red, rounded or kidney-shaped appendage. Sessile, shieldlike antheridial receptacles [fig. 24c] are commonly seen in summer. Female receptacles [fig. 24b] occur on separate plants in fall and winter as conic protuberances finally elevated in the spring on stalks having a single rhizoid furrow. There are 5–8 tubular involucres, each enclosing a single sporophyte, but no pseudoperianth is formed between involucre and calyptra. The pendent capsules dehisce by means of an apical cap and subsequent splitting into 4–8 longitudinal valves. The spores are multicellular at maturity (but the segmentation is difficult to discern because of exceedingly thin cell walls).

Conocephalum conicum (L.) Lindb. [figs. 24–26] is a broad, silvery-green ribbon that can be as much as 4 or rarely 8 inches long. It emits a spicy-sweet fragrance when crushed in a living state. (The smell has been characterized variously as resembling turpentine, varnish, or mushrooms and also as "catty," but to me it is a pleasant fragrance, unlike any of those odors.) The species is needlessly confused with *Marchantia*, from which it differs in its larger size, pale color, less noticeable midribs, coarsely sectored upper surface, and large, white stomatal structures. Also, the stomata are not surrounded by tiered guard cells, there are no gemmae, and the male and female receptacles are quite different. The bispiral elaters are often larger at 1 end, and the spores, produced in April and May, are 66–138 μm and densely papillose [fig. 25b–d].

The species' name is taken from the conic shape of the female receptacles [fig. 25a], and so is the generic name (meaning cone head).

The plants grow on seepy organic soil in mineral-rich swamps in Allegan, Barry, Berrien, Clare, Eaton, Gladwin, Gratiot, Ingham, Ionia, Isabella, Jackson, Kalamazoo, Lapeer, Lenawee, Livingston, Mecosta, Montcalm, Muskegon, Newaygo, Oakland, Oceana, Shiawassee, Tuscola, Van Buren, and Washtenaw Counties.

Sperm cells are said to be forcibly ejected from the antheridial receptacles in little puffs resembling jets of steam.

Indians of British Columbia once used these tonguelike plants in doctoring cankers of the mouth.

AYTONIACEAE

5. **Reboulia** Raddi

The thallus, of moderate size, is a forked ribbon of a light, dusty-green except for red margins and undersurfaces. The crenate margins are strongly incurved when dry. The upper surface is smooth (at least when dry), and its pores are quite inconspicuous, especially when dry. The pores are surrounded by 4–5 concentric rows of cells in 1 layer and scarcely elevated. The epidermal cells are thin-walled except for corner thickenings. The upper tissue consists of 1 layer of air chambers that are, however, incompletely partitioned and therefore seem to be several-layered. The air chambers lack green filaments. The ventral scales, imbricate in 1 row on each side of the midrib, are irregularly oblique-triangular with 2 linear appendages. The sessile, kidney-shaped, dusky-purple male receptacle is located just behind the stalked female receptacle, which is somewhat convex and broadly 4–7-lobed, though more commonly 5-lobed. The stalk has a single rhizoid furrow, and there are many slender, whitish scales at top and bottom as well as some rhizoids along the length of the stalk. Each sporophyte is enclosed in an involucre. There is no pseudoperianth. The spherical capsule dehisces by an irregular lid above a cuplike base.

The genus was named for a Florentine author of works on tulips, Eugène de Reboul, born in southeastern France, in Aix-en-Provence.

Reboulia hemisphaerica (L.) Raddi [figs. 27, 28] is easily confused with *Preissia* but has a smooth epidermis seemingly without polygonal areas and stomatal structures (at least when dry), crenate margins incurved when dry, slender, whitish scales at both ends of the stalks of female receptacles, divided air chambers, and epidermal cells with corner thickenings. The receptacles, instead of being squarish and 4-lobed, are most commonly 5-lobed. The spores (produced in April to June) are yellow, reticulate on the outer face, and winged at the junction of outer and inner faces.

The species' name refers, not too accurately, to the more or less half-spherical shape of the female receptacle.

In both *Reboulia* and *Conocephalum* the stomatal structures are simple except in the female receptacles where they are barrel-shaped.

The plants are suited to relatively dry habitats. During the heat of summer, the thallus protects itself from drying by curling up and exposing its scale-bearing undersurface. The plants are not obligate calciphiles, as *Preissia* is, and they characteristically grow on soil or rock on wooded ravine banks and hillside grasslands. They have been found in Barry, Eaton, Ingham, Kent, Livingston, Oakland, and Washtenaw Counties.

6. **Asterella** P.-Beauv.

The thalli, of moderate size, are divided into straplike forks. The epidermal pores are surrounded by 2–4 concentric series of cells in 1 layer; they are not elevated. The air chambers are, in most species, incompletely partitioned, and they lack green filaments. The ventral scales are arranged in 2 rows. The male receptacles are sessile but not well marked. The female receptacles, elevated on a slender stalk with a single rhizoid furrow, are hemispheric to conic, usually scarcely lobed, and bumpy above. On the undersurface are generally 4 bell-shaped, membranous involucres each enclosing a single archegonium (or occasionally a short row of them). After fertilization each archegonium comes to be surrounded by a pseudoperianth that splits into many long scales joined at the tips or free. The capsules are spherical and dehisce by means of a lid.

The generic name, meaning little star, presumably refers to the conspicuously dissected (but not actually starlike) pseudoperianths.

Asterella tenella (L.) P.-Beauv. [fig. 29] is relatively slender (often approaching a *Riccia* in size and appearance). It may be light green with reddish margins or uniformly dull red. The ventral scales are red. The plants are scarcely areolate and have inconspicuous pores. The epidermal cells lack corner thickenings. The air chambers are several-layered (but not divided by incomplete partitions). The female receptacles, elevated on wiry, dark-red stalks, are conic and slightly 4-lobed, red and rough-bulging above, but pale below. Each of the 4 capsules is enclosed in a large, reddish or straw-colored pseudoperianth that becomes deeply split into 6–10 incurved scales [fig. 29a–b]. The orange-yellow spores are 80–85 μm, areolate on the outer face, and broadly bordered. Spores in October and November.

Plants with fruiting structures have been found in Monroe County, on rain-washed sand of an old field near Bigelow, Raisinville Township. Sterile plants have been have been collected on wet, organically enriched sand along a power line in Gratiot Co. and also on alluvial soil of old fields in Monroe, Washtenaw, and Wayne Counties. (Elsewhere, the species is said to show a preference for relatively dry habitats.)

Asterella tenella is reported to be delightfully aromatic, but I have detected no odor. The specific epithet means slender, and the plants are indeed small as compared with *Preissia* or *Reboulia.*

Sterile thalli can be differentiated from those of a *Riccia* by the presence of red scales on the undersurface and at the notches marking the growing points.

MARCHANTIACEAE

7. **Preissia** Corda

The ribbonlike thallus, of moderate size, grows by dichotomous forks and also by ventral branches formed near the lobe tips. The color is a dusty gray-green except at purplish margins and undersurfaces. The reddish or blackish purple ventral scales are overlapped in 1 row on either side of the midrib. Semilunar in shape, they each bear a small, lanceolate appendage. The upper surface is minutely and inconspicuously areolate. The pores, elevated as tiny dots, are surrounded by several tiers of cells in a barrellike arrangement. Those cells are papillose on their exposed surfaces. Four cells of the lowest tier protrude in such a way as to form a crosslike opening. The air chambers are 1-layered and lack green filaments. The plants, nearly always dioicous, have male and female receptacles elevated on stalks with 2 rhizoid furrows. The male receptacle is a scarcely lobed disc on a short stalk. The female receptacle, with age usually maroon or brown, is squarish and shortly 4-lobed with 4 radiating ridges on its upper surface. Under each lobe an involucre surrounds an inflated pseudoperianth [fig. 25e] that in turn encloses 1–3 capsules. The capsules dehisce by 6 or 7 irregular valves.

The generic name does homage to Balthasar Preiss, an army physician and naturalist of Prague.

The barrellike stomatal structures of *Preissia* and *Marchantia* have an adjustable aperture (owing to changes in turgor of the four cells making up the bottom tier) and presumably serve the same function as the stomata of higher plants.

The absence of oil cells in the ventral scales is a unique feature among the Marchantiales of southern Michigan.

Preissia quadrata (Scop.) Nees [figs. 25e, 30], in contrast to *Reboulia,* is distinctly areolate with epidermal pores tiny but readily

apparent with a lens; the cells of the barrel-shaped stomatal apparatus are papillose; the epidermal cells lack corner thickenings; the
ventral scales have no oil cells; the female receptacle is only 4-lobed,
and the stalk lacks a scaly fringe at top and bottom. Although the
inflorescence is normally dioicous, fruiting structures are common.
(It is said that occasional monoicous plants are found.) The stalk of
the female receptacle can be as much as 10 cm long; that of the male
receptacle is much shorter. The spores are about 63 μm in diameter,
reticulate, and bluntly and unevenly lobed at the margins. Spores in
May and June.

The specific designation refers to squarish female receptacles.

The plants grow on wet, calcareous substrates, on soil, rock, and
logs in or near streams and ponds, often in rich fens, in Clare, Huron,
Oakland, and Van Buren Counties. Steere recorded occurrences in
Kalamazoo and Eaton Counties.

In the thickened midrib are developed brown, thick-walled fi-
berlike structures. As in *Conocephalum* and *Marchantia,* the cells
of the solid ventral tissue bear faint papillate markings on their
walls.

The plants are often peppery-hot to the taste, tingling the tip of
the tongue and burning the back of the throat. (The sensation is
similar to that caused by crystals of calcium oxalate produced in
jack-in-the-pulpit tubers.)

8. **Marchantia** L.

The thallus is a moderately large, dichotomously forked ribbon commonly bearing fringed gemmae cups that are ciliate-rimmed all
around [figs. 31, 35]. The pores are surrounded by cells in several
tiers, the 4 cells of the lowest tier bulging inward to form a crosslike
opening. The thallus surface is relatively smooth, but the polygonal
areas marking air chambers are readily discerned. The air chambers,
in 1 layer, are lined at bottom by short, green filaments. The brownish lower surface has an abundance of rhizoids as well as scales in 3
rows on either side of the median groove. The gemmae are discoid,
notched at either end, and attached by means of a short stalk. The
male and female receptacles, on separate plants, are elevated on
stalks. The male receptacle is a round disc with short lobes, and the
antheridia develop inside chambers below the dorsal surface [fig.

33b]. The female receptacle [fig. 32] is deeply radiate-lobed, and its stalk (actually a modified branch) has 2 rhizoid furrows. Groups of archegonia [fig. 33a] in 2 rows hang from the lower surface and alternate with the lobes. Two-lobed, ciliate involucres envelop each 2-rowed group of several archegonia, and small bell-shaped pseudoperianths enclose each sporophyte. The sporophyte [fig. 33c–d] consists of an ellipsoidal capsule, a short neck, and a massive foot embedded in gametophytic tissue.

The genus was named for Nicolas Marchant, director of the Duc d'Orléans' gardens at Blois. It was his son who bestowed that honor on him, and, in effect, on both of them. (The German botanist, Buxbaum, who in 1712 found the quaint moss that now bears his name, considered naming it for his father, as Marchant had done, but reminded himself of the fox that begged grapes, not for himself, but for his sick mother.)

Marchantia polymorpha L. [figs. 32–34] differs from *Conocephalum*, with which is it is commonly confused, in its smaller size, dark-green color, blackish midribs, smoother texture, less conspicuous stomatal structures, and almost constant presence of gemmae cups [fig. 31a]. Also both the male and female receptacles are stalked, and they are completely unlike those of *Conocephalum*. The spherical spores (produced from June to August) are yellow, smooth, and surprisingly small for the Marchantiales, only 10–14 μm in diameter.

The species is named for the many forms the plants assume.

Jerome Bock, who translated his name into Hieronymus Tragus, in his *De Stirpium Historia* of 1552, called *Marchantia* "a plant unfriendly to the summer sun, which delights in shady and likewise moist places and therefore grows only about the deeper and colder springs and on dewy rocks." *Marchantia* is actually a greenhouse weed that also grows outdoors in wet places, on soil, humus, and logs, especially where there has been a recent burn, but it disappears from sight as minerals released by fire leach away and larger plants get established. Collections have been made in Clare, Clinton, Gladwin, Huron, Isabella, Kalamazoo, Lapeer, Mecosta, Montcalm, Newaygo, Shiawassee, St. Clair, and Washtenaw Counties.

Marchantia's fine-grained texture contrasts with the alligator hide of *Conocephalum*.

On the walls of cells in the solid ventral tissue it is usually possible to discern a faint pattern of reticulated wall thickenings.

The pale and delicate ventral scales are arranged in three rows on either side of the midrib [fig. 34j]: those at the margins of the thallus (and projecting beyond the edge) are shortly lingulate; the intermediate scales are very narrow, elongate, and appressed to the thallus; and those along the midrib (but usually seen only near the lobe tips) are asymmetrically rounded-triangular and bear a reddish and roundish appendage. At the midrib is a cordlike strand of rhizoids enclosed in a tubelike covering of long, overlapping membranes. At either side of the midrib are rhizoids that penetrate the substrate. Parallel with the surface and covering the intermediate scales, strands of exposed rhizoids fan out from the median strand.

The antheridia develop inside chambers communicating with the dorsal surface of the receptacle in a succession such that the oldest are nearest the center and the youngest are at the margins, just back of the marginal notches. The archegonia are pendent in rows between rays of the female receptacle in order of maturity, the youngest being nearest the stalk. This arrangement appears to be just the opposite of that of the antheridia. It is actually the same, but a faster growth of the upper surface of the receptacle causes the tissues that lie between the lobes to grow downward and inward toward the stalk. The edge of the disc, in the sinus between the rays, is thus decurved, and the archegonia-bearing lower surface is really an extension of the upper surface.

The female receptacle [figs. 32, 34b, d] most commonly has nine rays, but owing to the position of the stalk, the involucres alternating with them are only eight. (In other species of the genus with differing numbers of rays, there is also one involucre fewer than the rays.)

The cells of the capsule wall are strengthened by annular thickenings. Cells internal to the wall become arranged in longitudinal rows [fig. 33c]. Some of them, usually after subdivision, become spores, as a result of meiotic divisions, and the others become long, slender elaters with spiral internal thickenings.

Because of its liverlike form (from which we derive the words liverwort and hepatic), *Marchantia* was once assumed to be useful in treating diseases of the liver. For that reason and many others, it was given prominence in the herbals of some centuries past:

"The decoction of Liverworte swageth the inflammation of the liver, openeth the stoppings of the same, and is very good agaynst fever tertians, and all inflammation of blood . . . . The same doth also heale all foule scurffes and spreading scabbes, as the Pockes, and wilde fire, and taketh away the markes and scarres made with hoate irons, if it is pounde with hony and layde thereupon. The same boyled in wine, and holden in the mouth, stoppeth the Catarrhes, that is, a distilling or falling downe of Reume, or water and flegme from the brayne to the throte."

—Dodoëns, 1578

I am told that in eastern Europe thalloid liverworts such as *Marchantia* floated in wine are considered a crunchy after-drink delight. (Would that not cause of a falling downe of Reume from the brayne to the throte?)

Order Metzgeriales

The thalli are ribbonlike or leafy. The rhizoids are smooth and never clustered (as in many leafy liverworts), and ventral scales are none. The thallus tissue is solid, and there are no epidermal pores. Leaflike lobes, if present, are in 2 rows. The cells never have thickened corners and often contain oil bodies (that disappear on drying). Sex organs are produced dorsally. Archegonia form behind the growing point, on the surface of the thallus or on short branches. The developing sporophytes are usually protected by an upgrowth of tissue (most often as a collarlike perigynium) or by a massive calyptra. The setae, elongating at maturity, are delicate and short-lived. The capsules have walls of some 2–5 layers of cells. They produce spores and spiral-thickened elaters and dehisce by means of 4 valves (or, in *Fossombronia*, by irregular rupture).

Many details of structure and developmental morphology are similar in the Metzgeriales and the Jungermanniales (often grouped together as the subclass Jungermanniidae). The similarity can be seen especially in the initiation and development of sex organs [figs. 9, 12] and the early development, structure, and dehiscence of the capsules.

In the Metzgeriales growth of the plant body, whether thallose or leafy, depends on a single apical cell, which is generally wedge-shaped, with two cutting faces, but rarely pyramidal, with three or four cutting faces (commonly known as 2-, 3-, or 4-sided cells according to the number of cutting faces). *Aneura, Riccardia, Metzgeria,* and *Blasia* have a 2-sided apical cell, and so does *Fossombronia,* our only genus with a leafy axis. In *Fossombronia* each segment gives rise to one leaf and a portion of the stem tissue. *Moerckia* has a 4-sided apical cell, and *Pellia epiphylla* has one with three cutting faces (but *P. calycina* has a 4-sided cell). Whatever the form of the apical cell, each lateral segment contributes to the thicker axis of the plant body and also to a continuous wing or a series of lobes or leaves.

Each cell cut off from the apical initial undergoes further divisions that result in cell blocks called segments or merophytes [figs. 7, 83a]. In the Metzgeriales, these merophytes give rise to lateral wings or leafy appendages toward the outside and the remainder of thallus tissue toward the inside. In the Jungermanniales (growing from a cell with three cutting faces), the outer cells of each merophyte contribute to the formation of leaves and adjoining cortex, while inner cells give origin to the medulla. Three rows of identical leaves so formed are considered to be the primitive condition, whereas a bilateral symmetry associated with two rows of lateral leaves and a third row of small underleaves, or none at all, is considered derived. A reduction of underleaves results when the ventral merophytes are down-sized or restricted in activity. The ventral merophytes can, in fact, be narrowed down to a single cell width.

Both thallose and leafy members of the Metzgeriales and also the Jungermanniales, all leafy, seem to have been derived from an erect, triradial, and leafless prototype. Evidence for such an ancestry is that nearly all liverworts of those orders grow by triradial segmentation of a tetrahedral apical cell. It is likely that early plants of erect growth and radial symmetry became prostrate in response to the rigors of life on land, in order to conserve water, increase photosynthetic exposure, and facilitate water-dependent fertilization. And leafy expansions of surface probably also developed in response to the conditions of terrestrial life, as they did in vascular plants of similar origins and early adaptive histories.

The Metzgeriales are often called the Anacrogynae because the archegonia are formed back from the apex [figs. 38a, 39d], behind the growing point (in contrast to the Jungermanniales, or Acrogynae). As in the Jungermanniales, the archegonia have five rows of neck cells, the spore mother cells become deeply 4-lobed before meiosis [figs. 37g, 40c, 49h], and the capsule wall may be 2–10 cells thick, with cell walls usually thin except for localized thickenings [figs. 14, 25d, 38d]. The patterns of such thickenings determine the mode and effectiveness of a dehiscence dependent on water loss and shrinkage. The epidermal cells are usually nodose-thickened, especially along their side walls. Cells internal to the epidermis have, most commonly, annular or semi-annular bands of internal thickening.

The Jungermanniidae never develop protective structures, such as perianths or perigynia, around single archegonia. (The Marchantiales, by contrast, commonly have pseudoperianths enclosing individual archegonia and involucres around single or grouped archegonia.)

The spores of the Metzgeriales, in contrast to those of the Jungermanniales, are typically large (mostly 35–100 μm) and sometimes several-celled, having germinated precociously [figs. 4f, 92d, 93h, 97f]. Smaller spores in the Jungermanniales seem to be associated with growth on tree trunks and faces of cliffs where wind is especially effective as an agent of long range dispersal.

1. Plants leafy.   6. *Fossombronia*
1. Plants thallose.
   2. Thallus without a midrib.
      3. Thallus small, much branched.   1. *Riccardia*
      3. Thallus of moderate size, scarcely branched.   2. *Aneura*
   2. Thallus with a thickened middle and broad, thin margins.
      4. Thallus broad in proportion to its length; midrib obscurely differentiated, gradually narrowed toward the margins, without laciniate scales on its upper surface.   3. *Pellia*
      4. Thallus elongate, strap-shaped; midrib abruptly differentiated and distinct, with laciniate scales (associated with sex organs) on the upper surface.
         5. Midrib with a small but distinct central strand; margins flat but ± wavy.   4. *Pallavicinia*
         5. Midrib without a central strand; margins erect and usually strongly crisped.   5. *Moerckia*

ANEURACEAE

1. **Riccardia** S. Gray

The small, much branched, dark-green thalli are often crowded together and erect-ascending. The thallus is, in section, elliptic to lens-shaped, with no midrib and little or no marginal thinness. Oil bodies are few or none. The lower region of the thallus is occupied by a mycorrhizal fungus. Clusters of 2-celled gemmae are produced within epidermal cells, particularly at branch tips. (The gemmae are produced internally by a rounding up of cell contents rather than by an external budding off of whole cells.) The antheridia occur in 2 rows on very small, oblong or oblong-linear branches. The archegonial branches, very short and fringed at their edges, arise in notches at the thallus margin; lateral in origin, they often come to be ventrally displaced. There is no perigynium, but a shoot calyptra enlarges as a fleshy, club-shaped, papillose structure (with walls 6–8 cells thick). Stout unispiral elaters are attached in tufts to elaterophore tissue at the tips of valves of the capsule. More slender elaters also lie free in the capsule.

Riccardia takes its name from the Florentine Vincento Riccardi, a donor to Micheli's *Nova Genera Plantarum*.

In *Riccardia*, as in *Aneura*, the archegonia are mingled with short, green cilia in marginal clusters. The fleshy, perigyniumlike structure enclosing the developing sporophyte is actually derived from the archegonial venter, together with stem tissue below it. Such a "shoot" calyptra can be expected to bear unused archegonia at its sides. It is eventually ruptured at its tip by the growth of the capsule and the elongation of the seta.

The genus differs from other members of the Metzgeriales in its palmate or pinnate branching.

As in *Aneura*, the lower region of the thallus may be occupied by a mycorrhizal endophyte.

1. Thallus irregularly, often palmately branched, with cells multistratose throughout (or occasionally unistratose in a single marginal row); oil bodies none.   1. *R. latifrons*
1. Thallus 2(3)-pinnate, with cells unistratose in 2–3 marginal rows; oil bodies large, 1–2 per cell.   2. *R. multifida*

1. *Riccardia latifrons* Lindb. [fig. 36] is freely and irregularly branched. The ultimate branches are about 0.5–1 mm wide and often ± broader at the tips. The thallus is multistratose throughout (or occasionally unistratose in 1 or 2 marginal rows of cells). Oil bodies are lacking. The plants are monoicous. Spores mature from May to August.

The species' name refers, inappropriately, to broad lobes of the thallus.

This little liverwort is common on wet, decorticated logs in hard-wood and white cedar swamps, in crowded, dark-green patches or in mixture with other bryophytes, in Barry, Eaton, Montcalm, Newaygo, Oakland, and Washtenaw Counties.

The plants, when growing, are faintly scented, somewhat minty smelling.

Riccardia palmata (Hedw.) Carruth. has been found on rotten logs in swamps in Newaygo County and, rarely, farther north. The plants are smaller than those of *R. latifrons,* palmately branched, with branches often nearly parallel, and dioicous. The branches are linear and scarcely broadened toward their tips. The outer cells are small, with thick, red-brown walls.

2. *Riccardia multifida* (L.) S. Gray [fig. 37a–d] is ± regularly 1–3-pinnate. The ultimate branches are not broadened toward their tips, and the marginal cells are unistratose in 2–3 rows. Large, ellipsoidal, finely segmented oil bodies are seen especially in the middle portions of living thalli, 1 or sometimes 2 per cell. The plants are monoicous, with archegonia and antheridia mingled together in marginal cavities. Spores in May and June.

The name of the species means much branched.

This species of wet logs and peaty substrates in wet, calcareous habitats, such as white cedar swamps, was included in Steere's flora, but no southern Michigan specimens have been seen.

2. **Aneura** Dum.

The thalli, of moderate size, are broadly strap-shaped and only sparsely branched. They are fleshy, though brittle, dark, greasy-green, 10–16 cells thick at the middle, and gradually tapered toward margins that are 1–3 layers in thickness. The ventral tissue is occupied by an endophytic mycorrhizal fungus. Oil bodies are numerous and small. The plants are dioicous. Sex organs are produced on the dorsal surface of very small lateral branches (similar to those of *Riccardia*). A fleshy, smooth, club-shaped shoot calyptra serves as a protective structure. Green cilia surround the base of the calyptra.

Stout elaters are attached in brushlike tufts at valve tips, and more slender elaters lie free among the spores.

The generic name signifies the absence of a nerve, or midrib.

Aneura pinguis (L.) Dum. [fig. 37e–g] grows scattered as dark, bluish green, greasy-seeming ribbons, much larger than the species of *Riccardia* and scarcely branched. It is found in wet, calcareous habitats, in rich fens and openings in white cedar swamps, in Clare, Eaton, Genesee, Ingham, Lapeer, Lenawee, Livingston, Montcalm, Oakland, St. Clair, Van Buren, Washtenaw, and Wayne Counties. It has been reported from Allegan County (Kauffman, 1915). Spores in April to June.

The name of the species refers to a fat or greasy appearance. The plants have the faint "minty" smell that characterizes many others of the Metzgeriales.

I have not seen the hairs that are so commonly figured as covering the calyptra.

PELLIACEAE

3. **Pellia** Raddi

The thalli are relatively broad, irregularly forked ribbons with an obscurely differentiated midrib gradually thinning out toward somewhat upturned margins. The upper surface is smooth and shiny. The cells contain numerous small, ellipsoid, roughened oil bodies. The antheridia are enclosed, singly, in small, green, pimplelike chambers grouped on the dorsal surface well behind the growing point. Groups of archegonia are produced near the growing point in the protection of a flaplike or cylindric involucre. The setae can be as much as 5 cm long. The spherical capsules produce large, ellipsoidal spores that are multicellular because of precocious germination. Stout elaters are attached in a brushlike tuft at the base of the capsule, and others, more slender, lie free among the spores.

The generic name, as well as the specific designation *Pellia fabbroniana*, was given by Raddi to honor a Florentine lawyer Leopoldo Pelli-Fabbroni, the son of Giovanni Fabbroni, for whom Raddi named the moss genus *Fabronia*. (In so doing, Raddi translated

Fabbroni—meaning Big Smith—back to a Latin form with a different spelling.)

The name *Pellia* fortunately has priority over *Papa* S. Gray. That name honors Giuseppi del Papa, to whom Micheli dedicated a genus of flowering plants, *Papea*. Dumortier understandably considered Papa a man rather than a plant and likewise rejected *Riccardius* and *Pallavicinius* as mere Latinized versions of men's names. But why the feminine versions, *Riccardia* and *Pallavicinia*, are acceptable as plant names is not clear.

The sexuality and inflorescence type, essential for the identification of species of *Pellia*, are all too often lacking in herbarium specimens, and in any case, dried material becomes black, brittle, and difficult to work with. The monoicous *Pellia epiphylla* is usually easy to name because of an involucral flap anterior to a group of green-pimpled antheridial chambers [figs. 38a–b, 39a, d]. Dioicous species are troublesome to identify unless their distinctive involucres are present, and so it is a good idea to search for involucres in the field rather than hope to find them in the limited material of specimens hastily collected.

The sunken position of each antheridium in its own chamber results from the upgrowth of thallus tissue around the antheridial initial (as in the Marchantiales).

On drying and dehiscence of the capsule, violent twisting motions of fixed basal elaters [figs. 38c, 40e, f] cause the mass of spores and the more slender and much more numerous free elaters to be loosened up. The free elaters continue by more gentle movements to fluff up the spores for effective dispersal into the air. Within the genus elaters may have two spiral thickenings or as many as six.

1. Plants monoicous, with antheridial chambers well behind a flaplike involucre.   1. *P. epiphylla*
1. Plants dioicous, the involucre a ringlike or cylindric structure.
    2. Involucre shortly cylindric, short on 1 side and crenate-lobed at the mouth; cells of the middle with pale thickened bands (best seen in longitudinal sections).   2. *P. neesiana*
    2. Involucre tall, suberect, and ciliate-fringed at the mouth; cells of the thickened middle of the thallus without thickened bands.   3. *P. megaspora*

1. *Pellia epiphylla* (L.) Corda [fig. 39] is shiny-green or sometimes tinged with red (especially in the autumn). In section, the thickened thallus middle shows an irregular lacing of pale, delicate bands over the cell walls. The bands run vertically and transversely (and are more easily seen in longitudinal than in transverse sections). The plants are monoicous. The antheridial chambers are scattered bumps on the upper surface of the thallus well behind an involucral flap that is close to the growing point. Spores in April and May.

The species got its name from the fact that the flaplike involucre projects from the surface of a leaflike thallus.

The plants grow on organic substrates in swamps and other wet habitats. At Grand Ledge, Eaton County, they have been seen on sandstone and mineral soil. Collections have been made in Barry, Eaton, Newaygo, Oakland, Van Buren, and Washtenaw Counties.

2. *Pellia neesiana* (Gottsche) Limpr. [fig. 40a–f] is deep-green but commonly blackish or dark-reddish along the middle. The thickened middle shows in section an irregular lacing of pale bands on cell walls. The plants are dioicous. The involucres are short, ringlike, nearly horizontal cylinders, especially short on the side toward the growing point, and have an irregularly lobed and crenate margin. The ellipsoid spores measure 98–118 μm in longer diameter.

The species takes its name from Christian Gottfried Daniel Nees von Esenbeck (1776–1858), professor of botany at Breslau.

The plants have been found at Grand Ledge, Eaton County, on sandy, marshy soil, and in Allegan County, on mucky humus near a stream.

3. *Pellia megaspora* Schust. [fig. 40g] is a shiny deep green. The cell walls of the thickened middle of the thallus show no bandlike thickenings. The plants are dioicous, and the perigynia are suberect, elongate cylinders that are irregularly ciliate-fringed at the mouth. The spores measure 100–120 μm in longer diameter.

This species, farther north, at least, is a decided calciphile (to be expected in association with *Thuja*); it has been found in southern Michigan in Newaygo and Van Buren Counties on the soil of creek banks in swampy places.

This liverwort has , until recently, been known in eastern North America as *P. endiviifolia* (Dicks.) Dum. or *P. fabbroniana* Raddi. That species (in Europe and the Pacific Northwest) has smaller spores and in the autumn becomes reddish and produces an abundance of short, deciduous, apical branches (resembling the crispiness of endive). The epithet *megaspora* refers to the large spores (relative to those of *P. endiviifolia*).

The absence of thickened strands on cell walls in the median region of the thallus can be used to identify sterile plants. If fruiting specimens are at hand, the long-cylindric perigynium of *P. megaspora* is unmistakable in comparison with the short and lopsided one of *P. neesiana*. In *P. megaspora* the inner layer of cells making up the capsule wall lack bandlike thickenings; the elaterophore filaments number about 100, are very slender (not unlike the elaters), and have 3–4 spiral thickenings. By contrast, *P. neesiana* and (*P. epiphylla* too) has inner cells of the capsule wall semi-annularly thickened and only about 30 short, stout, bispiral elaterophore filaments.

The plants, when fresh, have a slight, mintlike fragrance.

PALLAVICINIACEAE

4. **Pallavicinia** S. Gray

The little branched, straplike thalli are more or less tapered to a stipelike base. They are bright-green and have a well-marked midrib and unistratose, unlobed wings with somewhat wavy margins. The midrib includes a small central strand of elongate, thick-walled cells. The cells of the thallus apparently lack oil bodies. Male plants have a row of antheridia protected by overlapping, green, toothed scales on both sides of the midrib. Female plants, broader than the male, bear groups of archegonia encircled by green, laciniate scales at intervals on the midrib. A tubular perigynium, ciliate at its mouth, grows up around the developing sporophyte. The capsule is ± cylindric.

The genus gets its name from Lazarus Pallavicini, one-time archbishop of Genoa and patron of Micheli's.

Pallavicinia lyellii (Hook.) Carruth. [fig. 41] is easily recognized by its narrow, ribbonlike thalli with well-marked midribs. Spores are shed from April to June.

The plants grow on wet humus in hardwood swamps in Barry, Calhoun, Clinton, Jackson, Kalamazoo, Van Buren, and Washtenaw Counties.

The species bears the name of the geologist Charles Lyell, who first found it in England.

I have not confirmed that *Pallavicinia* has an odor (said to be like that of *Moerckia*).

5. **Moerckia** Gottsche

The thallus is a broadly strap-shaped, sparsely forked ribbon not noticeably narrowed at the base, with ± erect, wavy margins. The pale or dark, translucent-green plants have a well-marked midrib and broad wings several cell layers in thickness. There is no central strand. Male plants have antheridia on apical portions of the midrib covered over by laciniate scales. Female plants produce, on the midrib, well-spaced, oblong-cylindric perigynia that are dentate at the mouth and surrounded by laciniate scales.

Moerckia was named for a Danish lawyer, A. Moerch (not Moerck, as the generic name seems to indicate), involved in some way, perhaps as a patron, in the preparation of the *Flora Danica*.

Moerckia hibernica (Hook.) Gottsche [fig. 42], apparently including in its synonymy *M. flotowiana* (Nees) Schiffn., is a pale, green or yellow-green ribbon with a prominent midrib and slightly to distinctly wavy margins that are generally several cells thick. (Shaded plants may be dark-green and scarcely undulate.) According to Schuster, spores are shed in New York State in March and April.

The name of the species makes reference to a type locality in Ireland.

This species has been found in Livingston County, at Tiplady "Bog," near Hell, on the wet bases of *Potentilla fruticosa,* and in Lenawee County, at Ives Road Fen, with *Aneura* and the moss *Campylium stellatum*, both good indicators of calcium-rich fens. In northern Michigan, the plants sometimes grow in deep shade in abundantly calcareous *Thuja* swamps emergent from shallow vernal pools, sometimes associated with *Pellia.* They also grow in the open in mineral-rich sedge mats marginal to lakes and at the margins of beach pools.

Presumably, in *M. hibernica* the midrib is triangular in section and 8–16 cells deep, and the wings are plane or nearly so, whereas in *M. flotowiana* the midrib is more gradually merged with wavy-crisped wings, trapezoidal in section, and 16–25 cells in depth. The wings of *M. flotowiana* tend to be more crisped at the margins, but plants growing loosely in the deep shade of *Thuja* swamps may virtually lack undulations, whereas those of crowded growth in open fens are wavy. *Moerckia flotowiana* is said, in shaded places, to have two small and indistinct "conducting strands" in its thickened thallus middle. The cells of those strands, differentiated, if at all, only in having brownish walls, are not at all comparable to those of the well-formed strand seen in *Pallavicinia.*

The plants, when fresh, have a faintly minty fragrance.

CODONIACEAE

6. **Fossombronia** Raddi

The gametophyte is a small, prostrate, leafy-lobed plant of a pale-green color. The rhizoids are violet-red. The leaves or leaflike lobes are alternately arranged and succubously overlapped (from the stem tip downward). They are often crowded-erect and form a seemingly continuous ruffle. Very irregular in form, the leaves are variously lobed-dentate and, toward the base, 2 or more cells thick. There are no underleaves. The antheridia form before the archegonia. Thus, the archegonia are scattered over the dorsal surface of the stem some-what behind the growing point, with the antheridia grouped behind them on an older portion of the stem. The antheridia are naked or partially covered by bracts. After fertilization, a bowl-shaped perigynium ruffled at the edges grows up around the developing sporophyte. The globose capsule, elevated on a short, stout seta, ruptures irregularly or by 4 imperfect valves.

The genus takes its name from Vittorio Fossombroni, who served as the Tuscan Minister of Defense and Finance.

Fossombronia has a flat stem edged by overlapping leafy appendages, sometimes only imperfectly divided. The genus in some ways links the thallose Metzgeriales and leafy Jungermanniales, but dorsal sex organs indicate a closer tie to the Metzgeriales. The pale-green, ruffled leaves and red-violet rhizoids are generically distinctive, but the species are mainly differentiated by spore and elater characters which are not usually seen until September or even October and for a period of barely two weeks' duration.

1. Elaters short and stout, with ringlike or less often unispiral thickenings.
                                                                    1. *F. cristula*
1. Elaters slender, with bispiral thickenings.   2. *F. foveolata*

1. *Fossombronia cristula* Aust. [figs. 43, 44] produces brown, reticulate spores measuring 44–56 μm wide mingled with stout, bent or curved, often branched elaters having annular or loosely unispiral thickenings. Spores from September to November.

The name *cristula*, meaning little crest, refers to the ridges bordering the spores.

Fossombronia grows on wet, peaty sand (about pH 4.5), especially in areas of seasonal draw down, at the edges of lakes and ponds and on disturbed, silty soil of old fields, at Three Mile and Knickerbocker Lakes (Van Buren Co.), Ely Lake and the Allegan State Game Area (Allegan Co.), the Gratiot-Saginaw State Game Area (Gratiot Co.), Augusta Township (Washtenaw Co.), and London Township (Monroe Co.).

The elaters [fig. 43b] are inconspicuous and likely to be overlooked at less than high power of the microscope. They are small and delicate (28–58×6–18 μm), variably short, broad, bluntly tipped, and sometimes branched. They are typically thickened by 5–9 pale rings that are rarely joined to form the rudiments of a spiral. Those of our other species, *F. foveolata* [fig. 45], are long and slender (60–100×10 μm) and distinctly bispiral (although the spirals sometimes branch so that 3–4 spirals may be seen for a portion of the length).

2. *Fossombronia foveolata* Lindb. [figs. 45, 46] has slender, bispirally-thickened elaters rather than the fat, imperfect, boomerangs of *F. cristula*. The spores (produced in late August to November) are 38–55 μm wide.

On wet, organically enriched sand of a post-glacial lake bed in

Gratiot County, and on silty sand of old fields in a post-glacial lake plain in Washtenaw and Wayne Counties.

Fossombronia foveolata is said to be "strongly odorous."

Some authors consider *F. cristula* no more than an expression of *F. foveolata* with malformed elaters [compare figs. 43b and 45]. The spores are much the same, but the thalli may be somewhat smaller, with more crisped and pointy-lobed leaves, and the elaters are strikingly and, I think, significantly different. However, the two species do occupy similar habitats and sometimes grow intermingled.

Order Jungermanniales (Leafy Liverworts)

The plants are leafy and often branched, sometimes pinnately so. The rhizoids, often clustered at the base of underleaves or less commonly lateral leaves, are smooth and lack internal pegs. The leaves are typically set in 3 rows, but the third row, of underleaves, is variously reduced and sometimes lacking. The leaves, 1 cell thick and lacking a midrib, are often lobed or toothed. Their cells, often thickened at the corners, with trigones, generally contain oil bodies of characteristic numbers, shapes, sizes, and colors. (The oil bodies disappear as the plants dry out.) The antheridia, produced singly or in groups in the axils of ± modified leaves, or bracts, are globose and long-stalked. The archegonia are terminal on stems or branches. In most cases, they are surrounded by a perianth, enveloped in turn by an involucre consisting of two bracts and (often) a bracteole that are commonly joined together. The seta elongates at maturity of the capsule but is hyaline and delicate, soon withering. The capsules dehisce by means of 4 valves. The cells of the capsule wall, 2–8(10)-stratose, generally have nodose thickenings, and those of the inner layer are usually reinforced with semi-annular bands. The elaters are usually free.

Most liverworts are leafy and belong in the order Jungermanniales. Lateral leaves may be transversely [fig. 47a] or obliquely attached to the stem, and obliquely inserted leaves may overlap succubously (from the tip of the shoot downward) [fig. 47b] or incubously (from the base to the tip) [fig. 47c]. Relatively few genera have an incubous arrangement, and rather few of those have unlobed leaves. Some genera have leaves complicate-bilobed, with two lobes folded together [fig. 47d–e]; in most cases the dorsal and ventral lobes are markedly differentiated from one another, an extreme being seen in *Frullania* [fig. 47d], where the ventral lobe is much smaller and usually exceedingly concave so as to resemble an upside down cup.

A fundamental feature of the Jungermanniales is growth from an apical cell with three cutting faces [figs. 6, 7], two lateral and one ventral, from which derivative cells are cut off in a spiral sequence. (As an exception, one genus, *Pleurozia*, has an apical cell with only two cutting faces.) Each segment contributes to a leaf [fig. 6f] and one-third of the stem, and therefore the leaves are 3-ranked, but never displaced into a spiral phyllotaxy as in mosses. Most of the Jungermanniales are flat-growing and dorsiventrally organized, but *Anthelia* and *Herbertus* are considered primitive in having three ranks of leaves of similar size and shape. Both of those genera illustrate another primitive feature, namely equally bilobed leaves. In the Jungermanniales leaves may be unlobed, but they at least go through a primordial condition of lobing. In virtually all species a leaf primordium consists of two cells. Only one may be meristematic giving rise to an unlobed leaf; each of them may divide to give rise to four growing points and lobes; or both initials may become meristematic, one dividing to form two lobes of a 3-lobed leaf. The apical initial of each lobe undergoes only a few divisions, as basal cells soon become meristematic and contribute to the bulk of the leaf.

Each lobe of a primordial leaf is tipped by a small cell projecting as a "slime papilla," and an undivided leaf bears as a vestigial second lobe a persistent slime papilla. Slime papillae not only mark the position of leaf initials but also give evidence of rudimentary leaves and lobes as well as extremely reduced underleaves [fig. 8d–f]. Short-stalked slime papillae [figs. 5, 6j, 9d], or mucilage hairs, commonly occur in leaf axils and overarch growing tips and developing sex organs. Slime produced by these papillae may protect meristematic cells from dryness and provide the wetness needed for fertilization.

(The unlobed leaves of the Calobryales and the leaflike appendages of the Metzgeriales, often irregularly lobed-dentate, never develop from 2-lobed primordia.)

At least 12 branching types can be sorted into three categories according to origin:

1. "Terminal" branches are initiated superficially close behind the apical cell and replace one or both initials of an adjacent leaf. In the most common type, exemplified by *Frullania* [fig.6h], the entire ventral half of a lateral merophyte gives rise to a branch. The leaf formed immediately above such a branch lacks the normal product of that half segment. In *Frullania* the small, concave ventral lobe is the one lacking. In the *Microlejeunea* type [fig. 6i], a branch is initiated in the dorsal half of a lateral segment, and the dorsal half of the adjacent leaf is accordingly missing. In the *Acromastigum* type [fig. 6g], a branch develops in place of half of an underleaf. In *Zoopsis*, a branch wholly replaces a lateral leaf.
2. Some branches, also considered "terminal," originate close to the shoot apex from superficial cells just behind leaf primordia. Such branches, of the *Radula* and *Lejeunea* types, are associated with normal leaves.
3. And still others arise adventitiously far back of the shoot apex from already differentiated cells of the stem interior.

True dichotomies, formed by an apical cell dividing to form two cell initials separated by a group of inactive cells [fig. 6a–c], are not seen in the Jungermanniales, although in *Bazzania* branch initials develop very near the apical cell of main shoots with the result that equal forking is simulated [fig. 53a].

The apical cell of the stem or a specialized branch is eventually transformed into one or more archegonia, and as a consequence growth stops. For that reason the Jungermanniales are often referred to as the Acrogynae. Plants may be unisexual, or dioicous, or bisexual, or monoicous [fig. 48a]. In dioicous species, the female inflorescence may be at the tip of a main shoot or, depending on the species, at the tip of a branch. The male inflorescences form at or near the stem tip but may become intercalary as stem growth continues. In paroicous species, the antheridia occur directly below the female inflorescence. Autoicous liverworts have male and female inflorescences on separate branches of the same plants. (Inflorescences that are synoicous have the two kinds of sex organs mixed together, and those that are heteroicous have sex organs variously located on the same plant and in other plants of the same population. These two inflorescence conditions are more commonly seen in the mosses.)

As in the Metzgeriales, the archegonia have five rows of neck cells. The archegonia are generally protected by an involucre of modified leaves, the bracts and a bracteole, corresponding to two lateral leaves and an associated underleaf. (Underleaves and the equivalent bracteoles are sometimes lacking.) The modified leaves are usually larger, more erect than the stem leaves, and often connate at base. The bracts tend to be essentially leaflike when the vegetative leaves are unlobed but deeply lobed and incised when they are divided. Inside the involucral bracts a perianth usually provides an added protection for the archegonia and the developing sporophyte.

The perianth [figs. 8a, 48b–d] appears to be formed by an evolutionary coalescence. The union of two flat leaves and an underleaf may yield a perianth with three angles formed at the junction of leaf margins, as in *Cephalozia* and *Lophocolea* [fig. 73a–b] or a tubular perianth without angles, as in *Ptilidium* and *Blepharostoma* [fig. 49a, e]. If the leaves are keeled, the perianth may be 3-angled, as in *Porella*, *Lejeunea*, and *Frullania* [fig. 93b–c], resulting from the fusion of two leaves and an underleaf, or it may be 2-angled and dorsiventrally flattened if no underleaves are involved, as in *Scapania* and *Radula* [fig. 97b, g], the angles in either case corresponding to the folds or keels of ordinary leaves. A modification is seen where no underleaves are present and a laterally compressed perianth results from the fused margins of two lateral leaves, as in *Plagiochila* [fig. 88e–f].

Some few genera and species have upgrowths of stem tissue, or perigynia, surrounding the archegonia and developing sporophytes. In *Harpanthus*, some species of *Solenostoma*, and *Marsupella*, a tubular perigynium subtends the perianth [fig. 48c]. In *Geocalyx*, *Calypogeja*, and *Trichocolea*, a perianth is wanting, and a long, fleshy, pendent pouch, or marsupium [figs. 48d, 72d] is formed as a "shoot calyptra" made up of hollowed-out stem tissue. In *Trichocolea* the shoot calyptra is surmounted by a perigynium.

The embryo passes first through a linear series of four cells, whereas that of the Marchantiales generally develops as a ball of eight cells. In both cases there is an early differentiation of cells into an inner endothecium and an outer amphithecium [see fig. 13]. Sporogenous tissue develops from the endothecium and accordingly occupies the entire capsule except for the wall. (The early form of the embryo has a lesser evolutionary significance than was once thought owing to the discovery of some variation among the Marchantiales. In *Marchantia* and *Preissia,* for example, there is a preliminary octant stage, but in *Conocephalum, Reboulia,* and *Asterella,* the first divisions form a chain of four cells, as in leafy liverworts. Both linear and octant embryos are found within the genus *Riccia.*)

Spore mother cells (as also in the Metzgeriales) become deeply lobed before meiotic divisions take place [figs. 37g, 40c, 49h]. The spores, not or only imperfectly polar, lack conspicuous triradiate markings.

1. Archegonia produced on the dorsal surface of the stem; leaves in 2 rows, crowded together and appearing to form a ruffled thallus margin; rhizoids bright purplish red.
   Fossombronia, p. 49
1. Archegonia produced at tips of stems or branches; leaves in 2 or 3 rows (the third row as smaller and otherwise differentiated underleaves); rhizoids not purplish red (except in *Solenostoma crenuliforme*).
    2. Leaves complicate-bilobed (consisting of 2 lobes folded over against one another).
        3. Dorsal lobe smaller than the ventral.
            4. Dorsal lobe oblong (with parallel sides); perianth terete, plicate.
               23. *Diplophyllum*
            4. Dorsal lobe ovate or rounded; perianth flat, not plicate.   22. *Scapania*
        3. Dorsal lobe larger than the ventral.
            5. Underleaves none.
                6. Plants tiny; rhizoids attached to the stem; leaf cells unipapillose.   27. *Cololejeunea*
                6. Plants larger; rhizoids attached to ventral lobes; cells smooth.
                   28. *Radula*
            5. Underleaves present.
                7. Ventral lobes deeply concave, generally shaped like an inverted cup.   25. *Frullania*
                7. Ventral lobes not or moderately concave, not cuplike.
                    8. Plants robust, pinnately branched; underleaves not lobed.
                       24. *Porella*
                    8. Plants very small, not much branched; underleaves bilobed.
                       26. *Lejeunea*
    2. Leaves not complicate-bilobed.
        9. Leaves deeply dissected into threadlike segments.
            10. Plants minute, algalike, not or scarcely branched; leaves transversely inserted, dissected nearly to the base into 3–4 uniseriate threads.   1. *Blepharostoma*

10. Plants larger, not algalike, pinnately branched; leaves lobed and ciliate.

    11. Plants 1–2-pinnate, not frondose; leaves transversely inserted to incubous.   *2. Ptilidium*

    11. Plants 2–3-pinnate, frondose; leaves succubous.   *3. Trichocolea*

9. Leaves not dissected into threads.

    12. Leaves incubous (obliquely inserted and overlapped from base to tip of stem).

        13. Leaves rounded or ovate, entire or sometimes notched at the tip.   *6. Calypogeja*

        13. Leaves elongate, coarsely toothed or lobed.

            14. Plants robust, ± dichotomously forked; leaves asymmetric, coarsely 3-toothed at a broad apex.   *5. Bazzania*

            14. Plants relatively small, pinnately branched; leaves symmetric, divided 1/2 way down into 3–4 decurved, fingerlike lobes.   *4. Lepidozia*

    12. Leaves succubous or transversely inserted.

        15. Leaves succubous (overlapped from the stem tip downward or sometimes well spaced and not overlapped).

            16. Leaves with the dorsal margin rolled backward and continued down the stem as a decurrent fold.   *21. Plagiochila*

            16. Leaves without a decurrent dorsal fold.

                17. Leaves unlobed and entire or bidentate.

                    18. Underleaves deeply bilobed; leaves ± oblong, wide-spreading moist or dry.

                        19. Leaves entire, those below the stem tip (male bracts) with a very small, basal, pouchlike lobe.   *15. Chiloscyphus*

                        19. Leaves bidentate.

                            20. Underleaves with a supplementary tooth or lobe on 1 or both sides; leaf apex ± indented or bidentate.   *14. Lophocolea*

                            20. Underleaves without supplementary teeth or lobes; leaves evenly bidentate.   *13. Geocalyx*

                    18. Underleaves very small or none (not deeply bilobed); leaves round, wide-spreading when moist but mostly erect in 2 approximate rows when dry.

                      21. Plants robust, growing with *Sphagnum;* upper leaves pointed and gemmae-tipped; leaf cells and trigones very large.   *8. Mylia*

                      21. Plants small to medium-sized, not growing with *Sphagnum;* leaf tips rounded and lacking gemmae; cells and trigones smaller (or trigones none).

22. Stem tips often bearing clusters of gemmae. 20. *Odontoschisma*
22. Stem tips not gemmiparous.
  23. Perianths tapered to a narrow, ciliate mouth; involucral bracts ± ciliate below. 9. *Jamesoniella*
  23. Perianths not tapered at the mouth; involucre not ciliate.
    24. Perianth broadly cylindric, smooth. 10. *Jungermannia*
    24. Perianth fusiform, plicate. 11. *Solenostoma*
17. Leaves lobed.
  25. Leaves usually 4-lobed. 7. *Lophozia*, in part
  25. Leaves 2-lobed.
    26. Leaves with broad, blunt or rounded lobes separated by a narrow sinus. 19. *Cladopodiella*
    26. Leaves with acute lobes separated by a wide sinus.
      27. Cortical cells of stem large, thin-walled, and transparent. 17. *Cephalozia*
      27. Cortical cells of stem not large, thin-walled, and transparent. 7. *Lophozia*, in part
15. Leaves transversely inserted and ± erect.
  28. Upper leaves (and bracts) crowded together in a head, much lobed, toothed, and crisped. 7. *Lophozia*, in part
  28. Upper leaves bilobed, not crowded in a headlike cluster or crisped.
    29. Plants slender, delicately chainlike; leaves inflated., divided into long, incurved lobes. 18. *Nowellia*
    29. Plants not chainlike; leaves not inflated, with lobes less elongate, not incurved.
      30. Leaves bilobed 1/2 or more of their length, the lobes often divergent. 16. *Cephaloziella*
      30. Leaves more shortly bilobed.
        31. Plants minute; gemmae at stem tips, ellipsoid. 12. *Harpanthus*
        31. Plants small but not minute; gemmae at lobe tips, stellate-angled. 7. *Lophozia*, in part

PSEUDOLEPICOLEACEAE

1. **Blepharostoma** (Dum.) Dum.

Very small and slender, the plants are pure-green and not much branched. The leaves are almost transversely inserted and dissected nearly to the base into 3 or 4 cilia consisting of 8–12 cells in a single row. The underleaves consist of 2–3 cilia. Oil bodies are none. Usually monoicous, the plants produce involucral bracts larger than the leaves, with forked or spinose lobes 2–3 cells broad at base. The perianths are inflated-cylindric below, 3-angled above, and ciliate-fringed at the narrow mouth.

The generic name refers to the mouth of the perianth ciliate-fringed as an eyelid is with eyelashes.

Blepharostoma trichophyllum (L.) Dum. [fig. 49] can escape notice because of its small size and algalike appearance. Gemmae, sometimes produced in conidialike chains at the tips of upper leaves, are green, spherical, and smooth. Spores in May and June.

The plants have been found on wet, decorticated logs in swamps, often in mixture with other bryophytes, in Clare, Eaton, Mecosta, and Newaygo Counties. Steere (1934) reported the species from a swamp near Ann Arbor, Washtenaw County.

The specific epithet refers to hairy, or ciliate, leaves.

Blepharostoma resembles another small and algalike liverwort, *Kurzia setacea* (Spruce) Joerg. (also known as a *Microlepidozia*), found farther north in *Sphagnum* bogs. Its leaves are 3-lobed to within two or three cells of their insertions, and the lobes have cells in two to three rows except at their tips [see fig. 8b–d for a very similar species, *K. pauciflora* (syn. *K. sylvatica*), not known to occur in the state]. *Kurzia* is actually related to *Lepidozia* and *Bazzania* rather than to *Blepharostoma*. Curiously enough, the type species of the genus *Kurzia* was described as a red alga!

PTILIDIACEAE

2. **Ptilidium** Nees

These curiously attractive plants have 1- to 2-pinnate stems and incubously overlapped leaves deeply divided into 3 or 4 unequal lobes fringed by long, unbranched cilia. The leaf cells are somewhat to distinctly thickened at the corners. The underleaves are similar to lateral leaves, though somewhat smaller. The inflated, club-shaped perianths are ciliate-fringed at the mouth.

The generic name makes reference to feathery leaves.

1. Plants relatively small and freely branched, in flat, green, golden, or rusty-brown mats; stem leaves densely imbricated; leaves lobed about 3/4 down, the largest lobe 6–8 cells wide.   1. *P. pulcherrimum*
1. Plants robust, sparsely branched, ± erect and loosely tufted, red or red-brown; stem leaves distant, loosely imbricated; leaves lobed 1/2 down, the largest lobe 15–25 cells wide.   2. *P. ciliare*

1. *Ptilidium pulcherrimum* (Web.) Hampe [fig. 50a–d] is relatively small and grows in flat, green, golden, or rusty-brown mats difficult to remove from the substrate. The leaves are lobed about 3/4 down, and the broadest lobe is only 6–8 cells wide at base. The marginal cilia are relatively long, and the leaf cells have rather large corner thickenings. Spores in late April to August.

The species' name means very pretty.

The plants grow on the bases of trees and on stumps and logs, often in rather dry places, in Clare, Eaton, Genesee, Gladwin, Huron, Ingham, Ionia, Isabella, Kalamazoo, Lapeer, Livingston, Mecosta, Montcalm, Muskegon, Newaygo, Oakland, Saginaw, Sanilac, Van Buren, and Washtenaw Counties.

2. *Ptilidium ciliare* (L.) Hampe [fig. 50e–f] is a rather robust liverwort growing erect in deep, loose tufts (up to 3 inches high) that are pale- to rusty-brown or sometimes red. (The tufts are easily removed from the substrate.) The leaves are divided about 1/2 down and fringed with rather short cilia (fewer than in *P. pulcherrimum*). The largest lobe is 15–20 cells wide. The leaf cells have large corner thickenings. Spores (in the Upper Peninsula) in late June.

The name of the species makes reference to ciliate leaf margins.

The plants grow on soil and rock. Much less common in Michigan than *P. pulcherrimum*, this species has been found in southern Michigan only at Grand Ledge, Eaton County. Steere recorded an occurrence at Mud Lake in Washtenaw County, but the only specimen at the University of Michigan so labeled belongs to *P. pulcherrimum* instead.

Trichocoleaceae

3. **Trichocolea** Dum.

The plants are large, regularly 2–3-pinnate, and woolly in appearance owing to the fine dissection of leaves, first into 3–4 narrow lobes divided nearly to the insertion and then into many long, simple and branched cilia so crowded that the succubous arrangement and lobing of the leaves are difficult to ascertain. The leaf cells lack trigones, but the surface walls are finely striate-punctate. The underleaves are much like lateral leaves but somewhat smaller. The developing sporophyte is enclosed in a long, erect, cylindric or narrowly club-shaped, hairy "shoot calyptra," formed by stem tissue hollowed out by the growth of the foot and surmounted by an upgrowth of perigynium. The calyptra and perianth are much reduced.

The genus got its name from the hairy sheath, or shoot calyptra, surrounding the sporophyte during its development.

Trichocolea tomentella (Ehrh.) Dum. [fig. 51] is light-green or yellowish, rounded-frondose, and feathery soft. The stems measure up to about 12 cm in length. The leaves are divided into 3 or 4 main segments each about 3 cells wide at the base. The shoot calyptra is about 7 mm long, tawny in color and covered with an abundance of well-spaced branched hairs and, toward the apex, old archegonia. (Sporophytes are very rare. The only specimens with shoot calyptrae and sporophytes seen were collected in Bavaria and northern Italy in May. Schuster reported fruiting in New York and Minnesota in late April to late May.)

The name of this beautiful species refers to a fine, woolly appearance. (The epithet is a diminutive derived from *tomentum*, a material suitable for stuffing cushions, such as wool or feathers.)

Common in depressions in cedar swamps farther north, in our area the species has been found in swamps in Berrien, Clare, Isabella, Kalamazoo, Lapeer, Newaygo, and Washtenaw Counties. It was recorded, no doubt correctly, from Van Buren County by Kauffman (1915).

Lepidoziaceae

4. Lepidozia (Dum.) Dum.

These rather small, pinnately to bipinnately branched plants grow in closely interwoven, dull, gray-green mats. The branches diverge at right angles, and some branches at the stem tips are flagelliform, but no rootlike, ventral branches (as in *Bazzania*) are formed. The leaves are incubously arranged and divided about 1/2 down into 3 or 4 decurved, fingerlike lobes. The underleaves are similar, though smaller. Antheridia and archegonia are found on the same plants on separate, short, ventral branches. The perianths are bluntly 3-cornered above.

The generic name, meaning scaly branch, refers to crowded leaves and bracts of involucral branches.

Lepidozia reptans (Dum.) Dum. [fig. 52] has large, hyaline, irregularly toothed involucral bracts and large, hyaline perianths tapering to an angled, dentate mouth. Red spores are shed in May and June.

The species' name refers to a horizontal or creeping habit of growth.

The plants characterize very old, very rotten stumps in white cedar swamps in Michigan's North. They have been found, presumably in the same habitat, in southern Michigan, in Clare, Livingston, Montcalm, and Newaygo Counties. Steere's report (1934) from Washtenaw County, could not be verified.

5. **Bazzania** S. Gray

The plants are repeatedly forked almost dichotomously and also produce pale, rootlike, ventral branchlets with minute leaves. The incubously overlapped leaves are nearly flat and wide-spreading. They are asymmetric and coarsely 3-toothed at a broad apex. The much smaller underleaves are generally 4–5-toothed. The plants are dioicous, and both kinds of sexual branches are ventral in origin. The perianths are spindle-shaped and obscurely 3-cornered above.

The genus was named for Matteo Bazzani of Bologna, a professor of anatomy and one of Micheli's patrons.

Bazzania trilobata (L.) S. Gray [figs. 53, 54] is unmistakable because of its large size, deep-green color, forked branching, asymmetric, coarsely toothed leaves, and ventral, rootlike branchlets. The leaf cells are rounded and thin-walled except for large corner thickenings. Spores in May and, elsewhere in the East, well into the summer.

The plants favor cedar swamps in the north, where they most

commonly grow on the sides of very rotted stumps, but occur less commonly in the south and often in cedar swamps, in Clare, Jackson, Lapeer, Montcalm, Muskegon, Newaygo, Oakland, and Washtenaw Counties.

The species' name refers to 3-lobed (more precisely 3-dentate) leaves.

Bazzania trilobata is said to have the smell of sandalwood.

The species bears some resemblance, but scant relationship, to *Lophozia barbata*, which has 4-lobed, succubous leaves and very small, inconspicuous underleaves.

CALYPOGEJACEAE

6. **Calypogeja** Raddi

These small to medium-sized plants grow in thin, pale- or whitish green to yellowish, translucent mats. The sparsely branched stems commonly end in clusters of (1)2-celled, yellow-green gemmae. The leaves, overlapped from the base toward the tip of stems and branches, are asymmetrically oblong-elliptic to oblong-ovate, with a broadly rounded, entire apex or a somewhat pointed and often notched tip. The underleaves are rounded, truncate, or variously indented to bilobed, and rhizoids are restricted to the base of underleaves. The short male and female branches develop ventrally, in the axils of underleaves. There is no perianth. The sporophyte develops on a short branch that forms a fleshy, cylindric, pendent marsupium (consisting of hollowed-out stem tissue) that is sparsely rhizoidal and crowned with scales. The calyptra is united to the marsupium for most of its length. (The marsupium can often be demonstrated by dissecting away the substrate.) The 4 valves of the capsule are suberect and ± twisted after dehiscence.

The name of the genus, denoting an underground chalice, makes reference to the pouch in which the sporophyte develops [fig. 57c]. Derived by contracting the words *calyx* and *hypogaeus*, it was originally spelled *Calypogeja* for reasons that are orthographically defensible. That spelling seems to be nomenclaturally correct, and it is accordingly accepted, even through a change from a time-honored orthography more suited to the pronunciation, *Calypogeia*, seems tedious.

The leaves are unlobed and incubously overlapped, whereas all our other genera with unlobed leaves show a succubous arrangement. *Odontoschisma* has terminal clusters of yellowish gemmae and unlobed leaves, but the succubous arrangement of leaves provides an easy means of differentiation, and the leaves are also much firmer, not pale and translucent.

1. Underleaves rounded to truncate or somewhat retuse at the apex.
   3. *C. integristipula*
1. Underleaves distinctly bilobed.
   2. Leaves often bluntly pointed and sometimes notched at the apex; underleaves deeply bilobed and often notched or bluntly toothed at the sides. 1. *C. fissa*
   2. Leaves rounded at the apex, not notched; underleaves bluntly bilobed as much as 1/4 down, not toothed. 2. *C. muelleriana*

1. *Calypogeja fissa* (L.) Raddi [fig. 55a–d] has leaves somewhat longer than broad, bluntly or subacutely pointed, and often notched at the tip. The deeply bilobed underleaves are wider than long and generally bluntly toothed on 1 or both sides. Spores in May and June.

The specific epithet refers to split, or divided, underleaves.

The plants have been found in Clare, Eaton, Kalamazoo, Lapeer, Newaygo, and Van Buren Counties, on moist, shaded humus, sandy-organic soil, and rotten logs in swamps.

The actual separation of *C. fissa* from *C. muelleriana* is not clearly established. All North American collections have been referred to the subsp. *neogaea* Schust., characterized by oblong-ovate leaves and wide-spaced, broad underleaves that are bilobed nearly to the base and bluntly toothed or lobed at the sides. The leaf lobes are often bidentate on poorly developed plants, only rarely on mature, healthy shoots.

2. *Calypogeja muelleriana* (Schiffn.) K. Müll. [figs. 55e–g, 56] has broadly oblong leaves with narrowly rounded tips. The underleaves, broader than long and at least twice as wide as the stems, are bluntly bilobed as much as 1/4 down, with a triangular, often narrow sinus and normally entire margins.

The species was named for the distinguished German hepaticologist, Karl Müller.

The plants have been collected in Eaton, Isabella, Kalamazoo, Lapeer, Montcalm, Newaygo, and Oakland Counties.

This "species" is often impossible to differentiate from *C. fissa.*

Most American records of *C. trichomanis* (L.) Corda [fig. 58] belong here. Only a few North American collections have been made, in Maine, Vermont, Quebec, and Nova Scotia. The species is illustrated here because the name has been used so much in North America in the past and the species may indeed occur here. Unfortunately, the best character for recognizing the species–blue oil bodies–cannot be demonstrated in dried specimens. Because of the oil bodies [fig. 57], the plants, when living, have a bluish cast, especially in their younger parts. (The blueness is caused by a pair of sesquiterpenes appropriately named azulenes.)

3. *Calypogeja integristipula* Steph. [fig. 59a–d] (*C. meylanii* Buch) has broadly oblong-ovate leaves (longer than wide) rounded at the tips, and the underleaves are very large, 2–3 times wider than the stems, rounded, truncate, or emarginate and atypically very shortly bilobed at the apex. Spores in May and June.

The species' name makes reference to the essentially entire, undivided underleaves.

The plants commonly grow on wet humus or rotted stumps in swampy places, including cedar swamps. They have been found in Clare, Lapeer, Montcalm, Newaygo, and Tuscola Counties.

About four-fifths of all leafy liverworts are dioicous, and a good many of those produce gemmae. *Calypogeja integristipula* is monoicous but also reproduces itself by gemmae. Schuster (1983) reported seeing a patch no more than 4 inches square with some 50–100 sporophytes simultaneously producing perhaps 500–100 million spores! It is no surprise that such a species, whether dioicous or monoicous and fruiting or not, is often common and widely distributed. Of course, species reproducing by gemmae or by the fusion of gametes produced on the same plants are not likely to vary much from one population to another.

Michigan reports of *C. neesiana* (Mass. & Carest.) K. Müll. belong here. That species has leaves often somewhat narrowed to a shallow-notched tip, and in poorly developed plants the outermost row of cells may be more or less elongated parallel with the margin. (It appears that typical material of *C. neesiana* does not occur in North America.)

In *Sphagnum* bogs, one might expect to find, in association with *Mylia* and *Cladopodiella,* the tiny *Calypogeja sphagnicola* (Arn. & Perss.) Warnst. & Loeske [fig. 59e]. It is easily recognized, with a lens, by its widely spaced, asymmetric, more or less pointed leaves spreading at an angle of only about 45°. The underleaves are bifid to the middle or beyond.

JUNGERMANNIACEAE

7. **Lophozia** Dum.

The stems are creeping or ascending and sparsely branched with rhizoids distributed throughout. The leaves, obliquely attached and succubous to nearly transverse, are 2-lobed or, in some species, 2–5-lobed. The underleaves are generally small and obsolete to lacking. Gemmae, commonly at tips of leaf lobes, are usually angled and reddish. The female bracts are usually larger than the leaves and generally more lobed and dentate. The perianths are oblong-cylindric to obovoid and plicate above.

The generic name, meaning crested branch, was originally applied to species with an involucre made up of lobed bracts in contrast to the simple, unlobed bracts of *Aplozia.* The genus defies definition, and its taxonomy is notoriously troublesome. Mercifully, the species of southern Michigan are few and relatively easy to recognize.

1. Leaves 2-lobed; perianths smooth.
    2. Gemmae none; perianths inflated-clavate; leaf cells without trigones.
    4. *L. badensis*
    2. Gemmae present on deformed leaves of upcurved stem tips; perianths oblong-obovoid; leaf cells strongly trigonous.   5. *L. heterocolpos*
1. Leaves, at least at stem tips, normally more than 2-lobed; perianths plicate above.
    3. Leaves (as well as bracts) 2–4-lobed; gemmae not angled.   3. *L. capitata*
    3. Leaves of lower portion of stems 2-lobed; gemmae angled.
        4. Plants dioicous; female bracts and upper leaves capitate-crowded, many-lobed, and conspicuously crisped.   2. *L. incisa*
        4. Plants paroicous (the male bracts just below the female inflorescence); bracts and upper leaves less noticeably crowded, with lobes fewer (2–4), not particularly crisped.   1. *L. bicrenata*

1. *Lophozia bicrenata* (Hoffm.) Dum. [fig. 60] is very small, at most only about 5–6 mm long, green or red-yellow, and ascending at the tips. The transversely inserted leaves are crowded and suberect to somewhat secund and bilobed about 1/3 down. The leaf cells are thick-walled, especially at the corners, and therefore ± rounded, and the cuticle is granulose-roughened. The cells at lobe tips lack oil bodies and for that reason in fresh material appear to be dead. Underleaves are lacking. Pale-orange or red-yellow clusters of gemmae are borne at lobe tips of upper leaves which become much eroded and lacerate; the gemmae are 2-celled and bluntly angled to stellate. The male bracts are situated immediately below the female inflorescence or on separate plants in the axils of scarcely differentiated upper leaves. The involucral bracts and uppermost leaves are larger and more concave than the other leaves, 2–4-lobed, and spinose-dentate. The oblong-cylindric, inflated perianths are deeply plicate above and rather abruptly narrowed to a hyaline, ciliate-dentate mouth. According to Schuster, spores are produced throughout the growing season, from May to late October.

The species takes its name from its bluntly 2-lobed leaves.

The plants, decidedly xerophytic, grow on close-packed soil and partly covered by dry sand on 2-track roads through aspen stands and in lichen barrens occupying old burns in Michigan's north (often in company with *Cephaloziella rubella* in places where the Bug-on-a-Stick moss, *Buxbaumia*, grows). They have been found on rock in a sandstone quarry at Grand Ledge, Eaton County, and also in Huron County.

The small size, shallow bilobing of leaves, and pale-orange gemmae clustered at the tips of upper leaves aid in identification.

Fresh plants are said to smell of cedar oil, but I have been able to detect only a faintly minty odor.

Two collections from Grand Ledge, Eaton County, were named by Steere and confirmed by Miller as *Lophozia excisa* (Dicks.) Dum., but the specimens are much too poor to serve as vouchers for a species far disjunct from northern Michigan and adjacent Minnesota. *Lophozia excisa* is a small liverwort, pale-green in the shade, reddish in the sun. The leaves are transversely attached. The lower ones are squarrose, but the upper ones are capitate-crowded, erect-incurved, and wavy. They are shallowly bilobed, with relatively thin-walled cells. The male bracts, situated just below the female inflorescence, are irregularly 2–3-lobed and dentate. The perianths are broadly oblong and somewhat tapered to a plicate, irregularly toothed mouth. Gemmae, often borne at the tips of upper leaves, are red-brown or purplish, 1–2-celled, and angled.

2. *Lophozia incisa* (Schrad.) Dum. [fig. 61] is a small, pale, bright-green liverwort with ascending tips. The leaves are transversely inserted and erect-spreading. The lower leaves are bilobed, and the upper leaves, crowded together in a head, are unequally 2–5-lobed, with lobes acute and spinose-tipped, and have crisped, irregularly dentate to spinose margins. The leaf cells (in a living condition) have an abundance of small, roundish oil bodies (about 25–30 per cell). The rather thin cell walls have small but well-marked corner thickenings. Gemmae terminating the lobes of upper leaves are pale-green, 1–2-celled, and angled. The plants are dioicous with broadly obovoid perianths obtusely pleated above and abruptly narrowed to a ciliate mouth. Spores in May and June.

The species gets its name from its deep-cut leaves.

Rather common on rotten logs and peaty turf in the white cedar swamps of northern Michigan, the species has been found at Grand Ledge, Eaton County, and was reported from Port Austin, Huron County (Schnooberger, 1940).

The pale clusters of much lobed and crisped leaves resemble heads of crinkly lettuce. Plants that have not yet formed such heads can be recognized by their unequally 2-lobed leaves and, when fresh, whitish green color and large numbers of minute oil bodies. The leaves become dark-green on drying.

3. *Lophozia capitata* (Hook.) Macoun [fig. 62] is a rather small, pale-green liverwort (becoming red-tinged when growing in the sun). The leaves are obliquely inserted, but at the ascending stem tips they are very crowded and erect or incurved. They are much broader than long, irregularly 2–4-lobed, and ± wavy. The cells are relatively large (averaging more than 40 μm in diameter); they may be thin-walled and lacking in trigones or firm-walled and trigonous; they contain a large number of small oil bodies (16–60 per cell). Underleaves are absent. Gemmae are rarely produced in branched fascicles at the tips of normal leaves or in globular masses on much

reduced leaves at the tips of slender branches clustered near the stem apex; they are light-green, smoothly rounded-ovoid, and 1-celled. The plants are dioicous and have involucral bracts larger than the leaves, 3–5-lobed, ± dentate, and strongly crisped. The long-cylindric perianths are plicate and somewhat tapered to a ± ciliate mouth. (Both leaves and perianths tend to be reddish.) Spores in late April or early May.

The specific epithet refers to the headlike cluster of crowded leaves.

The species has been found on the wet bases of sedges at Grand Ledge, Eaton County. Elsewhere it grows on wet sand in ditches, sometimes, in the north, in jack pine peatlands.

Steere attributed *Lophozia barbata* (Schreb.) Dum., also known as *Barbilophozia barbata* (Schreb.) Loeske [fig. 63], to southern Michigan but gave no locality. It grows in northern Michigan on soil or humus in more or less shaded but usually rather dry habitats. The rather large and horizontal-growing plants are spicy-fragrant when alive and moist. The stems are felted with rhizoids on the undersurface. The deep-green or (when dry) brownish leaves, squarish in outline, are flat, wide-spreading, and obliquely inserted, with 3–5 broad, acute lobes (almost always 4). Very small, ciliate-lobed underleaves are restricted to stem tips. The plants are dioicous and have oblong-cylindric perianths plicate toward a crenate to dentate mouth.

4. *Lophozia badensis* (Gottsche ex Rabenh.) Schiffn. [fig. 64a–c] is a very small liverwort with creeping stems having hyaline, thin-walled cortical cells. The leaves, usually remote, are obliquely inserted, flat, wide-spreading, and uniformly 1/5–1/3-bilobed. The cells, delicate and thin-walled, virtually without trigones, are relatively large (up to 45×38 μm). The ellipsoid, nearly smooth oil bodies, 2–7 per cell, are very large. There are no underleaves. Gemmae are lacking. The plants are dioicous. The involucral bracts are erect and similar to the leaves but larger; they occasionally have a smaller third lobe. The perianths are smoothly inflated-claviform and abruptly contracted to a shortly beaked mouth that is crenulated by long cells. According to Schuster, spores are produced in New York State in the fall of the year, but in northern Michigan they have been seen in mid-August.

As its name indicates, the species was originally found in Baden, formerly a state of southwestern Germany.

The only collection known from southern Michigan came from Port Austin, Huron County (reported by Schnooberger, 1940, as *Cephalozia bicuspidata*). The species is to be expected in wet, calcareous soil, often in mixture with the tufa-forming moss *Gymnostomum*.

5. *Lophozia heterocolpos* (Thed.) Howe [fig. 64d–j] is very small and green, yellowish, or brownish, with erect-ascending stem tips. The leaves, obliquely inserted and widely spreading, are obtusely or sometimes acutely bilobed about 1/4 their lengths. The hexagonal leaf cells are smooth or faintly verruculose-striate and strongly trigonous, and they contain 3–6 golden-brown, ellipsoidal, finely segmented oil bodies. The underleaves, small and hidden among the rhizoids, are lanceolate and variously ciliate-dentate. Brown, smoothly ellipsoidal, 1–2-celled gemmae are borne on terete, upturned stem tips at the margins of erect-appressed, deformed leaves with oblong, thin-walled cells. The plants are dioicous, with involucral bracts leaflike but erect at base. The perianths are smoothly oblong-ovoid and abruptly contracted to a short beak.

On humus, problably overlying calcareous rock, at Port Austin, Huron County (*Schnooberger 1742*, MICH). This species of rare occurrence in the Upper Midwest has been found in Alpena and Presque Isle Counties in the Lower Peninsula and in Delta, Ontonagon, and Keweenaw Counties in the Upper, also in Minnesota, Wisconsin, and the Bruce Peninsula of Ontario.

The brownish, terete, upturned shoot tips bearing brown gemmae and strongly trigonous leaf cells are particularly diagnostic.

To be expected is *Anastrophyllum hellerianum* (Nees) Schust. [fig. 64k–o], a minute plant growing in sparse mixtures with other liverworts on decorticated, "ripe" logs in *Thuja* swamps, in especial association with *Nowellia*, *Jamesoniella*, and *Harpanthus*. The plants are decumbent but produce erect, terete, terminal shoots with much reduced branches bearing red gemma clusters; the leaves (much reduced on gemmiparous shoots) are bilobed and transversely inserted, and the gemmae are conspicuously red, angled, and unicellular. Plants have been seen from Iosco County, just north of the Tension Zone, and elsewhere to the north.

8. **Mylia** S. Gray

The plants are relatively robust. The undersides of stems are matted with rhizoids. The large, fleshy-seeming leaves are wide-spreading and succubously overlapped. They are rounded below, but toward the stem tips they become somewhat longer, ovate, and pointed. The large leaf cells have conspicuously bulging, yellow trigones. Small underleaves are hidden among the rhizoids. The plants are dioicous. The involucral bracts are similar to the leaves, and the wedge-shaped perianths are laterally compressed above and entire at the truncate mouth.

The genus takes its name from Willem Mylius, a Dutch physician who provided support for Micheli's *Nova Genera Plantarum*.

Mylia anomala (Hook.) S. Gray [fig. 65] is green or, in the sun, yellow to yellow-brown or rusty-red. The upper leaves commonly end in green or yellowish clusters of smooth, 2-celled, ellipsoid gemmae. The gemmae-bearing leaves are pointed and have elongate cells in contrast to the roundish cells of rounded lower leaves. The leaf cells are smooth, and the large, colorless oil bodies, 7–22 per cell, resemble clusters of grapes. Sporophytes are produced in August and September.

One of the few liverworts in Michigan restricted to peat bogs, *Mylia* grows in close association with *Sphagnum*, especially on the sides of hollows shaded by clumps of black spruce (in northern Michigan, at least, commonly in association with *Kurzia setacea, Cladopodiella fluitans,* and *Calypogeia sphagnicola*). Collections have been made in Jackson, Livingston, Mecosta, Montcalm, and Washtenaw Counties.

Mylia anomala is recognized by large, round to ovate leaves and yellow masses of gemmae on tips of upper leaves. It got its name from the "uncertain form of the leaves, varying even on the same individual, from orbicular to ovato-acuminate."

9. **Jamesoniella** (Spruce) Carring.

The plants, of moderate size, are creeping and sparsely branched with scattered rhizoids. The leaves, overlapping from the stem tips downward, are round to short-oval, broadly rounded or slightly indented at the apex, and entire. The underleaves are very small or none. The leaflike bracts surrounding the perianth are sparsely (and inconspicuously) ciliate toward their bases. The rather broadly cylindric perianths are tapered toward a plicate and ciliate mouth.

The genus was named for William Jameson, a Scottish botanist and physician living in Peru and later Ecuador. (In Quito he served as professor of botany and chemistry and also director of the mint.)

Jamesoniella autumnalis (DC.) Steph. [fig. 66] is dull dark-green and often tinged with red toward the apex (especially in the autumn on male inflorescences). The crowded and overlapping leaves spread almost horizontally when moist, but those toward the stem tips tend to be in upright rows when dry. The leaf shape is rounded-quadrate to very shortly oblong and rounded-truncate to emarginate at the apex. The leaf cells are irregularly rounded-polygonal and thick-walled with somewhat thickened corners. The plants are dioicous. The male inflorescence is a terminal, conelike group of 4–6 pairs of crowded, leaflike bracts. Spores in late summer and autumn.

The species was named in reference to fall fruiting.

Commonly covering the tops of rotten logs in swamps, the plants also grow on dry, sandy or humic soil of woodland banks; they have been found in Clare, Eaton, Huron, Isabella, Kalamazoo, Lapeer, Montcalm, Newaygo, Oakland, Van Buren, and Washtenaw Counties.

Unless they have developed a red tinge, sterile plants can be difficult to distinguish from others species of possible co-occurrence, such as *Jungermannia leiantha* and *Odontoschisma denudatum.* The more or less elongate, variably rectangular leaves of the *Jungermannia* stand in contrast to the rounded-quadrate ones of *Jamesoniella. Odontoschisma denudatum* has rounded, slightly concave leaves and stem tips ascending and terminated by clusters of gemmae.

Living plants have a slight fragrance. According to a Japanese worker, they often have a bitter taste that can be removed only by gargling. Many of the Jungermanniales and Metzgeriales have a medicinal flavor. I have noted this in many liverworts, perhaps most notably in *Ptilidium ciliare, Frullania eboracensis, Mylia anomala,* and *Tritomaria quinquedentata,* but I have not determined whether taste and smell are correlated. One might assume that taste and smell are associated with oil bodies, but *Radula complanata*, with its strikingly large oil bodies, has no taste or smell that I can detect. On the other hand, *Geocalyx graveolens* has a conspicuous development of oil bodies and a somewhat offensive odor. It seems that the various odors attributed to liverworts are caused, at least in part, by monoterpenes. Oil bodies consist largely of sesquiterpenes.

10. Jungermannia L.

The plants, of medium size, grow in light- or dark-green to brownish mats. The stems are creeping and only sparsely branched. The leaves, succubously overlapped, spread almost horizontally from the stem, except that the uppermost leaves and scarcely differentiated involucral bracts stand ± erect from a sheathing base. The leaves and bracts are oblong, rounded to truncate or somewhat retuse at the apex, dorsally decurrent, and entire. The rounded leaf cells, thickened at the corners but otherwise thin-walled, are often minutely verruculose and, toward the leaf base, ± striate. Underleaves are none. Gemmae, occasionally borne in short chains at the margins of small, deformed leaves at the tips of sterile shoots, are pale-green, round or ellipsoid, and 1–2-celled. The bracts of both male and female inflorescences are similar to leaves but erect-sheathing at base and ± spreading above. The perianths are smoothly inflated-cylindric, truncate at the apex, and abruptly narrowed to a short beak set in a shallow depression.

The generic name does honor to Ludwig Jungerman(n), an early teacher of botany and director of the botanical gardens at Giessen and Altdorf.

Jungermannia leiantha Grolle, also known as *J. lanceolata* L. [fig. 67], is recognized by shortly oblong, unlobed, flat leaves and fatly cylindric perianths. Leaflike male bracts subtend the female inflorescence. Spores in late April through May.

The plants occur on banks of soil in moist, shady ravines, on vertical rock surfaces, and also on rotten logs and humus in swamps in Barry, Berrien, Eaton, Huron, Kalamazoo, Mecosta, Newaygo, and Washtenaw Counties.

The species was named in reference to smooth perianths.

11. **Solenostoma** Mitt.

The plants are small and simple or somewhat branched with stem tips ascending but not gemmae-bearing. The rounded to elliptic, entire leaves overlap succubously but spread widely below and stand more nearly erect toward the stem tips. Underleaves are lacking. The involucral bracts are similar to the upper leaves but ± clasping at base. The perianths, broadly oblong-cylindric, ± spindle-shaped, or obovoid, are deeply plicate above. (In some of the species, a tubular extension of stem tissue, a perigynium, subtends the perianth.)

The generic name, meaning tubular mouth, makes reference, not too appropriately, to a cylindric perigynium present at the base of the perianth in some species and presumably resembling a razor clam, enclosing the archegonia and developing sporophytes. (*Solen*, the Greek name for a mussel of some sort, serves as the generic name for razor clams.)

Solenostoma is often included in *Jungermannia*, but the tapered, plicate perianths are significantly different, and so are the perigynia (collarlike upgrowths of stem tissue) subtending the perianths of some species of *Solenostoma*.

1. Leaves unbordered.   1. *S. pumilum*
1. Leaves bordered by a single row of swollen, thick-walled cells.
   2. Cells of leaf border up to 1.5 times as large as inner cells; rhizoids purplish red.   2. *S. crenuliforme*
   2. Cells of leaf border 2–3 times as large as inner cells; rhizoids colorless to brownish.   3. *S. gracillimum*

1. *Solenostoma pumilum* (With.) K. Müll. (*Jungermannia pumila* With.) [fig. 68] has dark-green or blackish, rounded-elliptic, unbordered leaves with cells thin-walled and smooth or nearly so. Male bracts are located immediately below the female inflorescence. The perianths are broadly tapered and plicate from the upper third to the apex, and there is no perigynium subtending the perianth.

The name of the species means small, dwarfish.

Plants have been found at Grand Ledge, Eaton County, on dripping cliffs of calcareous sandstone.

2. *Solenostoma crenuliforme* (Aust.) Steph. (*Jungermannia crenuliformis* Aust.) [fig. 69] is light- to yellow-green or, in the sun,

reddish. The rhizoids are purplish red. The rounded leaves are bordered by a single row of somewhat larger, thick-walled cells, and the margin is often upturned. The cell surfaces may be verruculose-roughened, and the inner cells are thick-walled and trigonous. The plants are dioicous, with an ovoid perianth and well-developed perigynium. Spores in May and June.

This species, reported by Steere as *Plectocolea hyalina* (Lyell) Mitt., has been found on moist ledges and cliff faces at Grand Ledge, Eaton County.

The specific epithet was chosen because of a resemblance to *Solenostoma crenulatum* Mitt. Concave leaves with a tumid border and purplish rhizoids identify *Solenostoma crenuliforme* in the field. (Among our liverworts only this and *Fossombronia* have purplish red rhizoids.) Under a dissecting microscope the swollen leaf margins are quite evident, but at higher magnifications the thickness is not easily perceived.

3. *Solenostoma gracillimum* (Smith) Schust. (*Jungermannia gracillima* Smith; *Solenostoma crenulatum* Mitt.) [fig. 70] is pale-green or, in the sun, pink-tinged. The rounded leaves are conspicuously bordered by a single row of thick-walled cells 2–3 times larger than the inner cells, which have thin walls and trigones none or small. The cells may be somewhat verruculose-roughened. The plants are dioicous, the perianths broadly ovoid, and the perigynia rather short.

The name given to this species means very slender.

This species has been found on clay banks and decayed sandstone at Grand Ledge, Eaton County.

GEOCALYCACEAE

12. **Harpanthus** Nees

The small, sparsely branched plants have shortly to deeply bilobed leaves that are succubous and spreading or nearly transverse and suberect. The underleaves are sometimes toothed on 1 or both sides and often united on 1 side to an adjacent leaf base. The leaf cells are thin-walled except for small corner thickenings. The plants are dioicous, and sex organs are produced on very short, ventral branches. The fruiting branch forms a hollowed out, fleshy perigynium below an oblong-oval to ovoid, tapered perianth.

The generic name refers to the perianth borne on a short, upcurved ventral branch supposedly resembling a sickle.

Harpanthus drummondii (Tayl.) Grolle [fig. 71] is minute and pale-green or yellowish with pale-green (or rarely red), 2-celled, ellipsoidal, smooth gemmae in clusters at ascending stem tips. The bilobed leaves may be suberect or, more commonly, flat and widespreading in 2 rows. The underleaves are ovate-lanceolate, usually entire-margined, and not at all or inconspicuously attached to the base of an adjacent leaf. The perigynium is scarcely as long as the involucral bracts, and the exserted perianth equals or somewhat exceeds it in length.

The species was named for Thomas Drummond, a Scotsman who collected widely in Canada and the eastern United States in the first half of the past century. (The type collection was collected by Drummond in "British North America.")

The plants grow on wet, decorticated logs in swamps in Barry, Ingham, Jackson, Kalamazoo, Newaygo, Van Buren, and Washtenaw Counties.

Our plants have long been called, erroneously, *H. scutatus* (Web. & Mohr) Spruce. The only expression of *H. drummondii* seen in southern Michigan resembles a *Cephaloziella* and is likely to be left uncollected because of the unsavory taxonomic reputation of that genus. (*Cephaloziella* has divergent leaf lobes and underleaves very small and often lacking.) This expression has been characterized by Schuster as "a persistently juvenile, sciophilous, androecial phase with abundant gemma formation on slight, subhyaline, pale to whitish-green shoots." (I have seen perianths only in a collection from Newaygo County, *Common 3221B*, MSU, as *Riccardia palmata*.) The typical form is stouter and has crowded leaves with incurved lobes and relatively large underleaves joined to the extreme base of an adjacent leaf. In our plants the underleaves are only rarely adnate to lateral leaves. (The reduced gemma-bearing plants can be confused with *Cephalozia* but can be separated from that genus rather easily by transversely inserted leaves and the possession of underleaves.)

13. **Geocalyx** Nees

The plants, moderate in size and sparsely branched, grow in flat mats of an opaque, bright-green or yellowish color. The leaves overlap from the stem tip downward and are horizontally spreading, nearly flat, and ± evenly bidentate at the apex. The large underleaves are deeply divided into 2 slender lobes. Rhizoids are mainly located near the base of underleaves. The short antheridial and archegonial branches are ventral in origin. There is no perianth, but archegonial branch tissue united with the calyptra forms a long, fleshy, pendent marsupium that is essentially smooth and only sparsely rhizoidal and has a few scales near its mouth. On dehiscence the valves of the capsule become spirally twisted.

The generic name makes reference to the underground marsupium as an earth cup.

Geocalyx graveolens (Schrad.) Nees [fig. 72] shows a neat regularity of wide-spreading, evenly bidentate leaves. The firm-textured and opaque, green or yellow leaves often contrast, when living, with a blue-gray stem. The leaf cells have 2–3 (or sometimes as many as 8–12) large and conspicuous, brown, elliptic, finely granulose oil bodies. Antheridia and archegonia are borne on separate branches of the same plants. The marsupium [fig. 48] is seldom seen. Spores in May and June.

The specific epithet gives notice of a strong and offensive smell. To me it is somewhat unpleasant but not at all turpentinelike as some authors say and certainly not pleasantly aromatic as Schuster says.

The plants grow on old logs and stumps and humic turf in swamps, both coniferous and hardwood, and also on dry banks of trails in Barry, Clare, Eaton, Ionia, Kalamazoo, Lapeer, Livingston, Mecosta, Montcalm, Newaygo, Oakland, Tuscola, and Washtenaw Counties.

Geocalyx is often confused with *Lophocolea heterophylla* and for no good reason: The leaves are firmer, more opaque, and consistently bidentate, and the deeply bilobed underleaves lack accessory lobes at their sides. When fresh, the leaves and stems, as seen under low power of a microscope, are conspicuously speckled with large, brown oil bodies.

Scanning electron micrographs of sporophytes within marsupia are displayed in *The Bryologist,* vol. 79, p. 268 (1976).

LOPHOCOLEACEAE

14. **Lophocolea** (Dum.) Dum.

Small to medium-sized, the plants grow in flat, pale, green or yellowish mats. The wide-spreading leaves overlap from the stem tip downward and are most often ± bilobed. Deeply bilobed underleaves bear on both sides an accessory lobe or coarse tooth, and rhizoids are restricted to the base of underleaves. The involucral bracts are coarsely toothed, and the bracteole is bifid and also toothed. The perianths, terete below and sharply 3-angled above, are deeply 3-lobed at a wide, coarse-toothed mouth. Two of the keels of the perianth are lateral, and one is antical.

The generic name, meaning crested sheath, refers to the perianth that is deeply lobed and toothed at the apex like a rooster's comb.

1. Plants of moderate size; leaves varying from bidentate to rounded or slightly indented at the apex; gemmae exceedingly rare.   1. *L. heterophylla*
1. Plants minute; leaves regularly bilobed, becoming ragged because of an abundance of marginal gemmae.   2. *L. minor*

1. *Lophocolea heterophylla* (Schrad.) Dum. [fig. 73], also known as *Chiloscyphus profundus* (Nees) Engel & Schust., forms thin, flat, translucently green or yellowish mats on old logs and also grows more crowded and ascending in firmer, darker-green tufts on the bases of trees and soil of wooded banks. The leaves vary considerably, even on the same plants, the lower leaves being bidentate at the apex and the upper ones rounded, emarginate, or only shallowly and bluntly bidentate. Sterile plants commonly have all the leaves bidentate. The leaf cells are thin-walled, with trigones minute or none. Gemmae are, very rarely, produced as short chains of 2–6 green, thin-walled, globose cells at the upper margins of leaves or on the surface of upper cells. The plants are monoicous (with the male bracts usually situated just below the female inflorescence). Spores are shed in May to July.

The plants are exceedingly common in moist woods and swamps, on the tops of decorticated logs, bark at base of trees, and disturbed soil of shaded roadbanks, in Allegan, Barry, Berrien, Clinton, Eaton, Genesee, Gratiot, Huron, Ingham, Ionia, Isabella, Jackson, Kalamazoo, Lapeer, Lenawee, Livingston, Mecosta, Montcalm, Muskegon, Newaygo, Oakland, Saginaw, Shiawassee, St. Clair, Tuscola, Van Buren, and Washtenaw Counties.

The name of the species emphasizes variations in leaf shapes. Plants growing in the sun are very different from those in the shade. Plants on rotten wood are usually larger than those on earth. Old plants have leaf shapes not seen in juveniles, yet a single tuft may show all gradations between juvenile and mature states. Rounded or truncate leaves tend to appear in the upper part of a fruiting stem and more bidentate leaves, representing a juvenile stage of growth, in lower portions [fig. 73c]. Colonizing plants may be very small, with very narrow lobes (about 2–3 cells in width), much resembling a *Cephaloziella*.

Sterile plants may key out in some manuals as *Lophocolea bidentata* (L.) Dum. That species, with consistently bidentate leaves and dioicous inflorescences, is very rare in eastern North America and unknown from Michigan.

Chiloscyphus has lateral leaves rounded or retuse at the apex, and the small inflated lobes at the base of leaflike male bracts are unmistakable. *Geocalyx* has firmer, more opaque, and consistently bidentate leaves, and its underleaves lack accessory teeth or lobes.

According to Schuster, many liverworts, including *Lophocolea heterophylla* and *L. minor*, have a rather distinctive "mossy odor." I have not detected any such odor (nor do I know what a mossy odor might be).

2. *Lophocolea minor* Nees [fig. 74], also known as *Chiloscyphus minor* (Nees) Engel & Schust., is a tiny, yellowish liverwort with rather distant, bilobed leaves bearing masses of 1-, 2-, or many-celled gemmae at their margins and as a result become ragged and eventually almost completely used up. Spores are dispersed early in July.

The species was named for its small size relative to *L. bidentata*.

The plants grow most commonly on disturbed, limy soil in mixture with mosses, especially on roadbanks and foot paths, in Huron, Mecosta, Newaygo, and Washtenaw Counties.

15. **Chiloscyphus** Corda

The plants are prostrate and sparsely branched, with leaves broadly attached, dorsally decurrent, and succubously overlapped. The leaves are broadly oblong, rounded, truncate, or emarginate at the apex, and entire at the margins. The leaf cells lack corner thickenings. The deeply bilobed underleaves bear on 1 or both sides at base an accessory lobe or tooth. Rhizoids are restricted to the base of underleaves. The plants are monoicous. Antheridia are produced, usually individually, under a small, dorsal, incurved-saccate lobe on leaflike bracts grouped in numerous pairs at the middle or near the tips of stems and branches. The bracts of the involucre are much smaller than the leaves and 2–3-lobed. The perianths, formed on very short branches, are 3-lobed at a wide mouth beyond which the calyptra is somewhat emergent.

The generic name refers to the perianth as a mouthlike cup.

1. Plants whitish green; median leaf cells 45–60 μm; perianth lobes irregularly spinose-dentate.   1. *C. pallescens*
1. Plants green, yellowish, or brownish green; median leaf cells about 35 μm; perianth lobes ± entire or bluntly toothed.   2. *C. polyanthos*

1. *Chiloscyphus pallescens* (Hoffm.) Dum. [figs. 75, 76a–c] has whitish green, translucent leaves that are somewhat longer than broad and rounded, truncate, or emarginate at the apex, with median cells 45–60 μm in diameter and somewhat thick-walled. The perianths are ± spinose-dentate at the mouth. Spores in May and June.

The species' name refers to a pale color.

The plants grow in swamps on moist soil and organic substrates, especially decorticated logs in Allegan, Clinton, Eaton, Ingham, Jackson, Kalamazoo, Mecosta, Newaygo, Oakland, Van Buren, and Washtenaw Counties.

Our species, *C. pallescens* and *C. polyanthos*, are commonly confused, but as the accompanying illustrations show [fig. 76], they differ to some extent in size and shape of leaves, nature of the leaf apex, size of leaf cells, thickness of cell walls, and toothing at the perianth mouth. (*Chiloscyphus polyanthos* has a chromosome number of n=9, whereas *C. pallescens*, with larger leaf cells, has n=18.)

2. *Chiloscyphus polyanthos* (L.) Corda [fig. 76d–f] has leaves about as broad as long and rounded at the apex. They have a pure-green or brownish color, median leaf cells only 25–37 μm and thin-walled, and perianth lobes subentire or bluntly toothed. Spores in May.

The species name gives promise of abundant "flowers," or perianths.

The plants have been found in swamps on decorticated logs and also on mineral or humic soil in Genesee, Gratiot, Isabella, Livingston, Oakland, and Washtenaw Counties.

Both of our species produce robust, dark or blackish green, aquatic expressions, known as *C. pallescens* var. *fragilis* (Roth) K. Müll. and *C. polyanthos* var. *rivularis* (Schrad.) Nees. Neither expression has been found in southern Michigan, but *C. pallescens* var. *fragilis* is rather common in swift-running brooks in the cedar swamps of northern Michigan. (Although hepaticologists seem to feel comfortable about downgrading these aquatic expressions to varietal levels, they are strikingly distinct in size and certainly in habitat from *C. polyanthos* and *C. pallescens*. Müller has said that *C. pallescens* var. *fragilis* passes into typical *C. pallescens* in culture.)

CEPHALOZIELLACEAE

16. **Cephaloziella** (Spruce) Steph.

These tiny plants are dark-green, brown, or reddish and little branched. The leaves, scarcely wider than the stems and transversely attached, are deeply bilobed; the lobes may be erect or divergent (but not connivent). Underleaves are very small or none. The perianths (with 3 main keels and several supplementary ones) are broadly oblong-cylindric to pear-shaped, deeply plicate toward the tip, and not much narrowed to a crenulate mouth.

The name is a diminutive for *Cephalozia*, a genus likewise characterized by deeply bilobed leaves, but in *Cephaloziella* the outer cells of the stem are not large and transparent, and the leaves generally have divergent, rather than connivent lobes.

The plants, almost too minute to be detected in the field, often grow in tufts of *Dicranum* and other mosses of dry, exposed habitats, such as *Cladonia* barrens. In such places, at least in northern Michigan, *Cephaloziella* often grows in blackish crusts on bare soil, often with *Lophozia bicrenata* and the moss *Buxbaumia aphylla.*

Sexuality can sometimes be determined in the field. In the case of a dioicous species, patches with and without perianths can be expected to occur separately. But, if most or all of the populations have perianths, it is reasonable to assume that the plants are monoicous. In herbarium specimens, if both male and female inflorescences are present in the same tuft, it is probable that the inflorescences were once joined to the same plants even though connections cannot always be demonstrated.

The minute, gemmae-bearing expression of *Harpanthus drummondii* is weaker and paler, with leaves less deeply divided into two non-divergent lobes.

1. Underleaves present throughout sterile stems; leaves often bearing a blunt
   tooth or short spine on 1 or both sides toward the base.   3. *C. spinigera*
1. Underleaves lacking on sterile stems; leaves entire.
   2. Leaf lobes, on sterile stems, 6–9 cells wide at base; leaf cells rather thin-
      walled.   1. *C. hampeana*
   2. Leaf lobes, on sterile stems, 3–5 cells wide; leaf cells thick-walled.
         2. *C. rubella*

1. *Cephaloziella hampeana* (Schrad.) Dum. [fig. 77] is dark- or brownish green and has rather distant, wide-spreading, even squarrose leaves with broadly triangular lobes that are distinctly divergent. The lobes are 8–12 cells long and 6–10 cells wide, and the cells are 11–18 μm wide and relatively thin-walled. The margins are entire. Very small underleaves occur near female inflorescences and near the tip of gemmae-bearing shoots. Gemmae, sometimes produced in apical clusters at the margins of reduced leaves, are pale green or slightly reddish, ellipsoid, smooth, and 1–2-celled. Antheridia and archegonia occur on separate branches of the same plants. The connate involucral bracts have 5–6 acute, dentate lobes, and the perianths are broadly oblong, wide-mouthed, and hyaline above. (The species, like *C. rubella*, fruits commonly and abundantly.)

The name does honor to the German bryologist who first found the species, Georg Ernst Ludwig Hampe.

This pygmy liverwort has been found on soil or soil over rock, presumably dry, once with the moss *Buxbaumia aphylla*, in Eaton,

Kalamazoo, and Montcalm Counties. Elsewhere, the species is usually found on moist or wet substrates, only rarely on dry, sterile sands.

2. *Cephaloziella rubella* (Nees) Warnst. [fig. 78] is dark- or red-brown. Its leaves have lobes not or somewhat divergent and only (2)3–5 cells wide. The leaves are entire, and the cells are about 12 μm wide and firm or ± thick-walled. Underleaves are lacking, at least on sterile stems. Gemmae, sometimes clustered at tips of stems at the margins of reduced leaves, are green or red, 1–2-celled, ellipsoid, and smooth. The sexual expression varies in the same populations, with archegonia and antheridia on separate branches of the same plants or with antheridial bracts subtending involucres. The connate bracts of the involucre have 5–6 acute, ± dentate lobes. The perianths, broadly oblong but somewhat narrowed to a wide mouth, are often purplish below and hyaline above. Spores have been observed in June and August.

The plants have been found on sandstone (Eaton County), on soil of a woodland bank with the moss *Dicranella heteromalla* and also on a decaying log in a hardwood swamp (Washtenaw County), on soil (Van Buren County), on dry sand with the mosses *Buxbaumia aphylla* and *Polytrichum piliferum* (Jackson, Livingston, and Muskegon Counties), and on *Sphagnum* in a *Chamaedaphne* bog (Montcalm County). I doubt if the habitat data of the latter can be trusted. Elsewhere in Michigan, this is a decided xerophyte characteristic of exposed, close-packed sand in pioneer habitats (commonly in association with *Lophozia bicrenata, Buxbaumia aphylla,* and species of *Cladonia*).

As indicated by the specific epithet, the plants become red (and also tend to have leaf cells more or less thickened) when growing in the sun. The leaves are much smaller than those of *C. hampeana,* with lobes much narrower.

A fruiting collection of the var. *sullivantii* (Aust.) K. Müll. from a rotten log in a hardwood swamp at Cedar Lake, Washtenaw County, shows most shoots toward their tips with coarsely toothed, bilobed leaves. The lower leaves are entire and narrow-lobed. The toothed leaves appear to be male bracts in several pairs on branches separate from the female inflorescence or on some shoots subtending it. (The male bracts may be quite entire or strongly toothed in this and the typical expression of *C. rubella*.) The variety, presumably found in moister habitats, has a suberect growth habit and perianths scarcely tapered to a wide mouth. It is generally found on old logs and peaty substrates in cedar swamps.

3. *Cephaloziella spinigera* (Lindb.) Joerg. (*C. subdentata* Warnst.) [fig. 79] is dark-green or somewhat tinged with red-brown. The epidermal cells of the stem may be smooth or faintly striolate. Minute underleaves of irregular form occur on sterile stems. The leaves are deeply divided into 2 divergent lobes that are about 3 or 4 cells wide. The margins sometimes bear a single blunt tooth or short spine on 1 or both sides below the middle. The leaf cells are about 17–25 μm long, with somewhat thickened, often brownish walls, and they sometimes bear low and inconspicuous papillae at back. Reddish tinged, ellipsoid, 1–2-celled gemmae are sometimes borne at stem tips. The plants are autoicous. The bilobed bracts and the bracteole are irregularly serrate, and the oblong-cylindric perianths are bluntly plicate above and crenulate at a wide mouth.

A few collections of sterile plants have been made in Livingston, Washtenaw, and Jackson Counties, in "boggy" habitats, in a rich fen and in a high bush blueberry "bog," clumped with *Sphagnum* and *Dicranum*. Farther north where bogs are better manifested, it seems to be associated with *Mylia*.

The name of the species, meaning spine-bearing, thorny, is not entirely appropriate inasmuch as the teeth at the leaf margins are few, as often as not lacking, and, in any case, scarcely thorny.

Cephaloziella elachista (Jack) Schiffn. (perhaps to be expected in our area) commonly has several sharp teeth at the leaf margins, and those on leaves and especially on bracts of the involucre are often recurved. It is, like *C. spinigera*, a bog species to be expected in association with *Mylia*.

CEPHALOZIACEAE

17. **Cephalozia** (Dum.) Dum.

These small plants of delicacy grow in flat, pale-green mats. The stems, not much branched, have an outer layer of large, thin-walled, transparent cells. The leaves overlap from the stem tips downward (but on sterile stems they may be widely spaced and not or only slightly overlapped). They are deeply bilobed and entire, and the lobes are generally separated by a wide sinus and often directed toward one another. Oil bodies are lacking. Underleaves are none or small and restricted to the vicinity of inflorescences. The male and female branches are ventral in origin, and the oblong-cylindric perianths are 3-angled above and somewhat tapered to the mouth. Two of the keels of the perianth are lateral; the third is postical.

The name of the genus, meaning headlike branch, refers to a crowding of leaves and bracts on short female branches.

The small size, deeply bilobed, transparent leaves, and large, transparent cells of the stem cortex are unmistakable. Though small, the plants are much larger than those of *Cephaloziella*, and the leaf lobes are never divergent. The genus is taxonomically difficult, and in many cases only perianth-bearing plants can be named with certainty. Our few species can generally be named in a sterile condition, but it is certainly reassuring to know the sexuality and perianth structure. By searching out perianth-bearing plants at the time of collection, one can avoid the taxonomic uncertainties that barren plants present.

1. Stems bearing slender, stoloniform or rootlike branches with minute leaves; leaves not decurrent, with lobes not connivent.
   2. Leaves symmetric, longer than broad, 1/2–2/3-lobed, the lobes sharp-pointed; mouth of perianth spiny-denticulate.   1. *C. bicuspidata*
   2. Leaves asymmetric, rounded, 1/3–1/2-lobed, the lobes often bluntly pointed; mouth of perianth shortly dentate.   2. *C. pleniceps*
1. Stems not bearing slender branches; leaves ± decurrent; leaf lobes often ± connivent.
   3. Plants dioicous; involucral bracts 2-lobed; mouth of perianth toothed; leaf cells 20–35 µm.   3. *C. lunulifolia*
   3. Plants monoicous; involucral bracts deeply 3–5-lobed; mouth of perianth ciliate; leaf cells 32–50 µm.   4. *C. connivens*

1. *Cephalozia bicuspidata* (L.) Dum. [fig. 80] is pale-green and often bears small-leaved, flagelliform branches. The leaves vary from moderately oblique to nearly transverse, and those near the stem tip are more nearly erect and crowded together. They are symmetric, longer than broad, lobed 1/2 or more their length, and non-

decurrent, and the lobes are slenderly pointed and suberect. The cells are longer than broad, about 38–50×28 μm, and thin-walled. The male and female inflorescences are on separate branches of the same plants. The female bracts are 2-lobed and entire or sparsely toothed, and the perianths are shortly and irregularly spiny-dentate at the mouth. The spores may be distinctly red or red-brown. Spores in May and June.

The plants have been found on wet soil over rock and humus in Clinton, Eaton, and Kalamazoo Counties. A record from Huron County (Schnooberger, 1940) belongs to *Lophozia badensis* instead.

The specific epithet refers to 2-pointed leaves. The leaves are nearly symmetric, with straight lobes and no decurrency.

2. *Cephalozia pleniceps* (Aust.) Lindb. [fig. 81] is a deep-green plant producing pale, slender branches from the underside of stems. (These branches are likely to be broken off when stems are pulled out for microscopic examination.) The broadly ovate to orbicular leaves are very obliquely inserted, wide-spreading, and concave. They are not or only slightly decurrent, and the lobes, not or slightly connivent, are unequal in size and often bluntly pointed. The median cells are thin-walled and very large, measuring 50–60×33–38 μm in diameter. Greenish, ellipsoid gemmae are sometimes clustered at tips of short, erect branches. Male and female inflorescences occur on separate branches of the same stems. The involucral bracts are 2–4-parted, and the perianths are plicate above and crenate at the mouth.

The name of the species apparently refers to the headlike fullness of clusters of gemmae.

The plants, growing especially on wet humus and rotten wood in swamps, have been found in Clinton, Eaton, and Washtenaw Counties.

The leaves are asymmetric but not or scarcely decurrent. The lower half of the perianth is bistratose and therefore firm and smooth, but the upper half is unistratose and plicate.

3. *Cephalozia lunulifolia* (Dum.) Dum. [figs. 82, 83b, at right], more commonly known as *C. media* Lindb., is pale green or yellowish and has no flagelliform branches. The nearly round, though asym-

metric leaves have distinct decurrencies and acute, usually connivent lobes (1- or 2-celled at the tips). They are obliquely inserted,
wide-spreading, and nearly flat. The median leaf cells are only about
23–35 μm and thin-walled. Pale-green, 1-celled, angular gemmae are
frequently clustered at stem tips. The plants are dioicous, with female bracts 2-lobed and entire and with perianths short-toothed at
the mouth. Spores in May to July.

The specific epithet refers to crescent-shaped leaves.

This species is found on wet organic soil and turf and also on rotten
logs and stumps in swamps, both coniferous and hardwood, commonly in association with *Sphagnum*. It is known from Clare, Gratiot, Jackson, Mecosta, Montcalm, Newaygo, and Oakland Counties.

Cephalozia lunulifolia, probably our most common species, has
asymmetric, decurrent leaves like those of *C. connivens*, but the
shortly toothed perianths are distinctive, as are also the smaller leaf
cells [fig. 83b].

4. *Cephalozia connivens* (Dicks.) Lindb. [fig. 83a–b, at left] is pale
and whitish green with no flagelliform branches and has nearly
round, decurrent leaves with sharp, somewhat connivent lobes. The
cells are firm-, though rather thin-walled, and large (about 40–60 μm
in diameter). Gemmae are rare. Male and female inflorescences are
produced on different branches of the same stems. The female bracts
are divided into 3–5 lanceolate segments, and the perianths are
ciliate-fringed at the mouth. The spores are dull red or red-brown.
Spores in May to July.

The species' name makes reference to leaf lobes directed toward
one another.

This species of swampy or boggy habitats is found especially in
association with *Sphagnum* in Calhoun, Jackson, Kalamazoo,
Lapeer, Livingston, Montcalm, and Oakland Counties.

Cephalozia loitlesbergeri Schiffn. was attributed by Steere to "the
northern part of our range." This species, restricted to *Sphagnum*
bogs, has leaves longer than broad, with sharp lobes connivent and
often crossing one another and leaf cells relatively small (28–35
μm). As in *C. connivens*, the plants are autoicous, the perianths are
ciliate-fringed at the mouth, and the female bracts are laciniate-
lobed.

18. **Nowellia** Mitt.

Small and delicate, this liverwort is clear-green or, in sunlight and especially in the fall of the year, red. The scarcely branched, prostrate stems are made chainlike by incurved leaf lobes. The outer cells of the stem are large and pellucid, though thick-walled. The leaves are transversely attached by a narrow base and asymmetrically divided into 2 long, slender lobes. The ventral lobe is semicordate at base and inflated. The leaf cells are evenly thick-walled and lack oil bodies. Underleaves are lacking. Gemmae, produced on the lobes of immature leaves of somewhat ascending branches, are 1-celled, pale yellow-green, and subspherical to ellipsoidal. The involucral bracts are 1/3-bilobed and dentate to spinose at the margins. The perianths, usually red below and hyaline above, are oblong-cylindric, sharply 3-angled, and ciliate at a broadly truncate mouth.

The genus was named for John Nowell (1802–67), of Yorkshire, a sweat-shop factory worker who apparently learned to read and write as a young man attending Sunday School. He worked long hours at a handloom, like Silas Marner, yet found time to become a respected and locally influential botanist.

Nowellia curvifolia (Dicks.) Mitt. [fig. 84] is appropriately named for inflated leaves with incurved lobes that give the stems the beauty of a delicate chain. The plants are nearly always dioicous, and the male plants are somewhat more slender than the female. Spores in May and June.

These small plants grow on decorticated logs in both coniferous and hardwood swamps, often in mixture with other liverworts including *Jamesoniella autumnalis* and (elsewhere in the state) *Anastrophyllum hellerianum*. Collections have been made in Berrien, Ingham, Jackson, Kalamazoo, Macomb, Montcalm, Newaygo, Oakland, St. Clair, Van Buren, and Washtenaw Counties.

19. **Cladopodiella** Buch

The slender, dark-green to blackish plants are sparsely branched except for ventral, flagellate branches with leaves reduced or lacking. The cortical cells of the stems are opaque and undifferentiated from the inner cells. The leaves, flat, obliquely inserted, and succubous, though seldom overlapped, are divided as much as 0.4 down into 2 broad, somewhat unequal lobes that are blunt or rounded at the apex and separated by a narrow sinus. The oil bodies are 2–10 per cell, small, globose or ellipsoidal, and very finely granulose. The underleaves, always present near the stem tip, are very small and irregular in shape (often ± bidentate). The plants are dioicous. The spicate male inflorescence consists of some 4–10 pairs of close-set, concave, bilobed leaflike bracts. The archegonia are produced on short, ventral branches, and the long-terete perianths are 3-angled above and somewhat crenulate at a wide mouth.

The name is a diminutive of *Cladopus*, a subgenus of *Cephalozia* once in use. Taken from Greek words for branch and foot, it refers to the short branch on the lower ("foot") side on which the sporophyte forms. The genus differs significantly from *Cephalozia* in having dark stems not corticated by transparent cells, dark leaves with broad, blunt lobes separated by a narrow sinus, and underleaves present, however small and inconspicuous.

Cladopodiella fluitans (Nees) Joerg. [fig. 85] grows among *Sphagnum* in soppy-wet bog habitats, often in puddles of water in soaks and drainage channels, such as deer trails. It also grows in bog hollows beneath clumps of black spruce. It has been found at Grand Ledge in Eaton County on peat at a pool margin and at Brighton Bog in Livingston County on the *Sphagnum* mat at the water's edge.

The name of the species means floating.

ADELANTHACEAE

20. **Odontoschisma** (Dum.) Dum.

The plants, of moderate size, have creeping stems ascending at the tips and few, ventral branches. Rootlike flagellate branches are produced toward the stem base, and rhizoids are scattered. The crowded leaves are obliquely to almost longitudinally inserted and succubously imbricated. They are slightly concave, rounded (or somewhat indented at the apex), and entire. The cell walls are thickened at the corners. The large, elliptic oil bodies are hyaline, finely segmented (like a bunch of grapes), and 2–7 per cell. Underleaves, generally small or obsolete, bear numerous slime papillae at the margins. The large perianth is 3-angled above and narrowed to a toothed or ciliate mouth; it forms on a short ventral shoot.

The generic name makes reference to the mouth of the perianth split into teeth (or cilia).

Odontoschisma denudatum (Mart.) Dum. [fig. 86] is green or ± red-brown. The stems generally turn up at the tip where the leaves are reduced in size and clusters of yellowish or pale-green, 1–2-celled gemmae are produced. Toward the stem base the leaves are also reduced. The crowded and overlapped leaves are rounded and somewhat saucerlike-concave. On drying they become ascending in 2 rows. The leaf cells are somewhat roughened, and the trigones are large. The underleaves are small and often worn away below, but toward the stem tip they are somewhat larger and commonly bifid at the tips. The plants are dioicous. The narrowly fusiform perianths, produced laterally, are dull-reddish and fleshy below but hyaline in the upper half, somewhat plicate above, and denticulate at the mouth. Spores in May to July.

The plants grow on wet humus, rotten stumps, and tops of wet, decorticated logs in swamps in Jackson, Kalamazoo, Newaygo, Oakland, and Washtenaw Counties.

Odontoschisma denudatum is easily separated from other entire-leaved species, such as *Jamesoniella* and *Jungermannia*, by conspicuously erect, gemmiparous shoot tips. The leaves are somewhat concave and diminish in size toward both ends of the shoots giving them a somewhat naked appearance that accounts for the specific epithet. Unlike *Calypogeja*, which has unlobed leaves and often terminal clusters of gemmae as well, *Odontoschisma* has succubous leaves.

PLAGIOCHILACEAE

21. **Plagiochila** (Dum.) Dum.

These moderately robust, loosely tufted plants have erect-ascending leafy shoots, almost lacking rhizoids, arising from a creeping, densely rhizoidal, leafy stem. The leaves, overlapping from the stem tip downward, are asymmetrically rounded to oval. The leaves are rolled when dry, and the dorsal margins, when moist, are rolled backward and conspicuously continued down the stem as a convex fold. Underleaves are very small or none. The plants are dioicous, with involucral bracts similar to leaves but somewhat larger and more dentate. The perianths are inflated below but laterally compressed toward the truncate apex.

The generic name, meaning oblique lip, bears reference to the fact that the flattened perianth mouth is sometimes bent. The genus is readily recognized, not by that feature of accidental occurrence, but by rolled leaf margins continued down the stem as a dorsal decurrency.

Plagiochila porelloides (Torr.) Lindenb. [figs. 87, 88], more commonly known as *P. asplenioides* (L.) Dum., has dark stems and leaves nearly straight at the dorsal margin and rounded at the ventral. The ventral margin may be strongly toothed, with teeth of 1–3 or even 5 cells, but the dorsal margin is usually ± entire and often the leaves are entire throughout. Underleaves are minute, acutely pointed structures found, if at all, especially near stem tips. Gemmae are lacking. The male inflorescence is a compact spike produced terminally but sometimes becoming intercalary as growth continues. The male bracts, imbricated in several pairs, are ventricose at base and spreading at their broad tips. The perianth is shortly ciliate at a broadly truncate mouth. Spores in late April to June.

The name of the species means, uninformatively or perhaps even inanely, *Porella*-like.

This liverwort grows at the sides of turfy depressions under the exposed roots of *Thuja* in cedar swamps and also on soil at the base of hardwoods in moist woods and swamps. It is known from Allegan, Clare, Clinton, Eaton, Huron, Ingham, Livingston, Newaygo, Oakland, and Washtenaw Counties.

SCAPANIACEAE

22. Scapania (Dum.) Dum.

Small to moderately robust and erect-ascending, the plants are sparsely branched and have leaves conspicuously complicate-bilobed and usually toothed at the margins. The smaller dorsal lobe is ovate to round and sharply folded over against the broad ventral lobe. Underleaves are lacking. The dorsiventrally compressed perianths are squared off at a broad mouth.

The genus got its name from its flat and supposedly hoe-shaped perianth.

The leaves have ventral lobes about as broad as long, whereas those of *Diplophyllum* are clearly elongate, and the perianths are flattened rather than fatly terete and plicate.

1. Plants relatively robust; leaves keeled, spinose-dentate; gemmae cinnamon-colored, 1-celled.   1. *S. nemorea*
1. Plants very small; leaves not or obscurely keeled, irregularly denticulate to dentate (especially on the dorsal lobe); gemmae reddish, 1–2-celled.
                                                                         2. *S. saxicola*

1. *Scapania nemorea* (L.) Grolle [fig. 89a–d], more commonly known as *S. nemorosa* (L.) Dum., is moderately robust. Shiny, deep-green and often red-tinged, the leaves have oblong-curved ventral lobes much larger than the arched, reniform or cordate dorsal lobes. The dorsal lobe is generally pointed, but the ventral lobe is rounded at the apex. Both lobes are decurrent, especially the ventral one. The margins are coarsely spinose, and the leaf cells have somewhat thick-ened corners. The oil bodies, 3–6 per cell, are large, pale brown, spherical to ellipsoidal, and finely segmented. Clusters of unicellu-lar, smoothly ellipsoidal, cinnamon-brown gemmae are often seen at tips of leaf lobes. The plants are dioicous, and the perianths are generally ciliate-dentate at the mouth. Spores in May and June.

The specific epithet refers to an occurrence in groves of trees.

The plants have been found at Grand Ledge, Eaton County, on wet sandstone blocks and ledges, and at the Detroit Metropolitan Airport in Wayne County, on the banks of an abandoned borrow pit. (Farther north, they occur on wet, peaty turf and old logs in white cedar swamps.) Nichols & Steere (1936) reported the species from the high banks of the Kalamazoo River near Battle Creek in Calhoun County.

Like most dioicous species, *Scapania nemorea* does not often fruit, but it commonly forms sizable populations as a result of asex-ual reproduction. Schuster (1957) has estimated that a foot-square patch of the liverwort can produce as many as one million gemmae in a single year!

The long-standing names of several common species, including this one so long known as *S. nemorosa*, have recently been altered or replaced for tiresome legalistic reasons. Such changes may be necessary, yet we all like the familiar. Peter Collinson said it for us long ago, in a letter to Linnaeus: "Thus botany, which was a pleasant study . . . is now become, by alterations and new names, the study of a man's life, and none now but real professors can pretend to attain it."

2. *Scapania saxicola* Schust. [fig. 89e–h] is a very small liverwort with reddish, 1–2-celled, smooth gemmae clustered at the tips of uppermost leaves. The leaves are not or only obscurely keeled, and the margins, especially of the dorsal lobes, are often irregularly dentate. The rounded-hexagonal leaf cells are only moderately thickened at the corners.

A very few plants were found in mixture with *Lophozia heterocolpos* in a scant collection from Port Austin, Huron County (*Schnooberger 1742*, MICH), on humus and presumably over a calcareous substrate.

The very small size, reddish gemmae, and scarcely keeled leaves with irregularly toothed margins give meaning to a species otherwise known from a few counties in Wisconsin and Minnesota.

23. **Diplophyllum** (Dum.) Dum.

The plants are small and slender with crowded and overlapping, complicate-bilobed leaves. Both lobes are elongate and oblong. The smaller dorsal lobe is folded over against the ventral one and directed forward at a 45° angle. The ventral lobe, spreading almost horizontally from the stem, at almost a 90° angle, is curved-asymmetric. Both lobes are finely and irregularly toothed all around. Underleaves are lacking. The perianths are terete, oblong-cylindric and deeply plicate.

The genus is named for leaves that appear to be doubled.

Diplophyllum apiculatum (Evans) Steph. [fig. 90] has ventral leaf lobes abruptly and conspicuously apiculate at an obtuse to rounded apex. The leaf cells, except in a few marginal rows, are obscurely roughened by 2–8 or more papillae. Numerous oil bodies resembling clusters of grapes nearly fill the cells and cause them to be opaque. Tiny, 1–2-celled, green, stellate gemmae are commonly clustered at tips of uppermost leaves. Antheridia are produced on a separate branch from the female involucre. Spores in May and June.

The specific epithet refers to abruptly pointed leaves.

This species is known in our area only from thin soil overlying sandstone at Grand Ledge, Eaton County.

In dorsal view, this small, rigid-seeming liverwort appears to have leaves in four flat rows.

PORELLACEAE

24. **Porella** L.

The plants are large and regularly 2–3-pinnately branched. The leaves are incubous and deeply complicate-bilobed. The larger dorsal lobes are broadly oblong. The ventral lobes are narrowly lingulate to rectangular, rounded, or ovate and parallel the stem or diverge slightly from it. The broad underleaves are commonly decurrent. The plants are dioicous, with shortly spicate male inflorescences at ends of short branches and perianths inflated-ovoid, obscurely 3-angled above, and toothed at the mouth. The capsules are spherical and, on dehiscence, deeply divided into about 8 segments. The large spores are multicellular at maturity owing to precocious germination.

The name *Porella* was used by Dillenius, who in his *Historia Muscorum* of 1743 erroneously described and illustrated capsules with numerous small pores at their sides. (His description concerned the liverwort, from Pennsylvania, that Linnaeus supplied with the binomial *P. pinnata*. Unfortunately, Linnaeus ranked *Porella* with the mosses.)

Some of the Porellas are reputed to have a "pungent" flavor, but I have detected only a slight to moderate taste in *P. platyphylla* and *P. platyphylloidea*.

The female inflorescence [fig. 91a] is produced at the tips of short branches. The male inflorescence [fig. 91b] is a short, spikelike branch, with antheridia axillary to numerous suberect antheridial bracts; the jacket cells are in 2 layers [fig. 91d]. As in other members of the order Jungermanniales, *Porella* has a capsule wall of several layers [fig. 91d].

1. Plants submerged or stranded; underleaves not broader than the stems, not decurrent.   3. *P. pinnata*
1. Plants never submerged; underleaves broader than the stems, long-decurrent.
    2. Ventral lobe of leaves about half as broad as the underleaves, oblong-ovate and concave-pointed; dorsal lobe longer than broad. 1. *P. platyphylla*
    2. Ventral lobe of leaves about as broad as the underleaves, almost circular, rounded at the apex; dorsal lobe ± rounded.   2. *P. platyphylloidea*

1. *Porella platyphylla* (L.) Pfeiff. [fig. 92a–d] is a large, conspicuously 2–3-pinnate liverwort of a dark- or olive-green to brown color. The leaves, distinctly overlapped and imbricated, impart a braided appearance to the upper side of stems and branches, but the ventral view of lateral leaves, underleaves, and lobules gives an impression of 5 rows of leaves. The dorsal lobe is longer than wide, oblong-ovate, rounded at the apex, and deflexed at the margins. The ventral lobe is about half as broad as the underleaves, oblong-ovate, bluntly obtuse, and somewhat concave, with the outer margin broadly recurved. The squarish underleaves, broader than the stems, are truncate to rounded at the apex and long-decurrent. The perianth mouth is beset with rather few (15–20) cilia and teeth of irregular shape and form. The finely papillose spores measure 55×36–40 μm, according to Schuster, but those I have seen are rounded and only 36–38 μm; the septa are difficult to discern owing to thin walls. The elaters are bispirally marked by thickened bands. Spores in May and June.

The name of the species means flat leaf.

This species is common on bark at the base of hardwoods, soil of steep, shaded roadbanks, and rock faces in Allegan, Barry, Genesee, Gratiot, Huron, Ingham, Ionia, Isabella, Jackson, Kalamazoo, Lapeer, Livingston, Mecosta, Montcalm, Newaygo, Oakland, Tuscola, Van Buren, and Washtenaw Counties.

Evans (1916), who made a careful analysis of the problems of intergradation exhibited by *P. platyphylla* and *P. platyphylloidea*, placed greatest reliance on the nature of elaters. In *P. platyphylla* they are bispirally thickened to their very ends where they are curiously continuous as loops. In those of *P. platyphylloidea*, on the other hand, the bands of thickening are unispiral (though occasionally doubled at the middle).

2. *Porella platyphylloidea* (Schwein.) Lindb. [fig. 92e–h], perhaps better known as *P. platyphylla* var. *platyphylloidea* (Schwein.) Frye & Clark, generally 1-pinnate in its branching habit, has ventral lobes about as wide as the underleaves, almost circular, rounded at the apex, and narrowly recurved at the margins. The larger, dorsal lobes are also nearly circular, and they have a small, crisped auricle on 1 side near their bases (where the lobe and lobule form a keel), and the underleaves are rounded and conspicuously decurrent. The perianth mouth is crowded-ciliate. The spores measure 40–45×36–37 μm, and the elaters are unispiral. Spores in August and September in Michigan (but said to be May and June elsewhere).

I have seen only a few collections from Southern Lower Michigan (all sterile), from Clare, Livingston, Lapeer, and Midland Counties, that I am willing to assign to this species, appropriately named for its troublesome resemblance to *Porella platyphylla*. Identification, though difficult at best, is simplified in Michigan by the fact that *P. platyphylla* is much more common than *P. platyphylloidea*. (Steere referred all material from southern Michigan to *P. platyphylloidea*, but his picture is an excellent representation of *P. platyphylla*.) *Porella platyphylla* is said to be distinctly less common in the eastern United States than *P. platyphylloidea* but almost as frequent in Minnesota (Schuster, 1953) and Kansas (McGregor, 1955). The width and shape of ventral leaf lobes are variable. McGregor has said that only juvenile plants, in Kansas, show signs of intergradation, but I am not prepared to agree, and Evans (1916) also found intergradations in the vegetative characters of even well-developed plants. Reproductive features are presumed to be more reliable as taxonomic characters. The perianths of *P. platyphylla* are only distantly toothed at the mouth, and the elaters are bispiral, whereas the perianths of *P. platyphylloidea* are crowded-ciliate, and the elaters are unispiral. Unfortunately, however, neither species fruits very often.

According to Schuster, making *P. platyphylloidea* a variety of *P. platyphylla* is unacceptable, because the two sometimes grow together and yet remain fully distinct. (Having been perplexed by the problem for half a century, I wonder if they might also grow together without retaining their distinctness.) The shape of the small ventral lobes of stem leaves and their width relative to the underleaves can be used in separating two entities pretty well, tongue in cheek, but Evans rightly cautioned against relying too heavily on such vegetative features, and Schuster went so far as to say, "Since *Porella* is so abundant, such specimens, representing intergradations in vegetative characters, should not be named *unless elaters can be studied.*" Of course, the difficulties of using sporophytic characters rarely seen in no way detracts from their importance, but throwing a troublesome sterile specimen out is no more of a solution than putting a pillow over one's head rather than facing up to the noise downstairs. Downgrading a species to a variety doesn't make the problem go away. Since dealing with such unsortables is and always will be frustrating, lumping them together may, after all, provide a better solution.

3. *Porella pinnata* Lindb. [fig. 92i] grows in loose, dark-green or blackish patches. The leaves, not or only slightly overlapped, are nearly flat when moist. The large dorsal lobe is asymmetrically oblong (with the upper edge arched, the lower straight) and rounded at the apex. The ventral lobe is no wider than the stem or even narrower, flat or nearly so, narrowly oblong and rounded at the tip, and nearly parallel with the stem. The underleaves, also no wider than the stem, are subquadrate, truncate or rounded at the apex, and not decurrent. The perianth mouth is shortly and densely ciliate (with cilia 1–4 cells long). [The spores are said to be 30–42 μm and the elaters 2-spiral or "locally" 3–4-spiral.]

The specific epithet means featherlike.

The plants grow attached to logs or brush and submerged in moving water or, more typically, in streamside areas subject to flooding, stranded in black masses during most of the year on the bases of trees and stumps. The species is known from river bottoms in Berrien, Genesee, Montcalm, and Shiawassee Counties.

JUBULACEAE

25. **Frullania** Raddi

Forming dark tracings on tree trunks and rock faces, the plants are loosely pinnate-branched with incubously overlapped, complicate-bilobed leaves. The dorsal lobe is rounded to ovate, rounded at the apex, and entire. The much smaller ventral lobe, or lobule, is usually very concave to the extent of resembling an inverted cup; it is connected to the dorsal lobe rather than the stem by a minute stalk. Near the base of the lobule is a minute, usually subulate appendage, or stylus. The leaf cells are generally thickened at the corners and often in between. The underleaves are bilobed. The sex organs form on short branches. The male branches have few to many pairs of leaflike bracts crowded together in a ± flattened spike. The bracts of the involucre are in 3–6 pairs and generally dentate to laciniate. The perianths, often dorsiventrally flattened, are obovoid to obcordate, bluntly 3-angled (with 1 ventral and 2 lateral keels and sometimes supplementary ridges too), and abruptly narrowed to a short beak. The capsules are spherical.

Leonardo Frullani, for whom the genus was named, was a Florentine privy counselor and director of the Tuscan treasury.

As in *Lejeunea* and *Cololejeunea*, the elaters are vertically oriented and attached at both ends. On dehiscence of the capsule, the flat lower ends become free, while the upper ends remain attached to the valves.

No poisonous substances are known to be produced by liverworts or mosses. However, in British Columbia, skin rashes affecting forestry personnel have been traced to some liverworts (and lichens), particularly to species of *Frullania* and *Radula* and attributed to certain sesquiterpenes. (Five of Michigan's *Frullania* species have been suspected of causing allergies: *F. asagrayana*, *F. bolanderi*, *F. eboracensis*, *F. inflata*, and *F. riparia*.)

1. Dorsal lobe of leaf with a median row of short, reddish cells. 5. *F. asagrayana*
1. Dorsal lobe without a median row of reddish cells.
  2. Autoicous (male and female inflorescences on different branches of the same plant); leaf cells usually without intermediate thickenings; lobules often explanate. 3. *F. inflata*
  2. Dioicous; leaf cells with numerous or occasional intermediate thickenings; lobules inflated or both inflated and explanate.
    3. Lobules mostly explanate; underleaves broad (as much as 3 or 4 times the stem width). 4. *F. riparia*
    3. Lobules inflated and cup-shaped; underleaves narrower.
      4. Lobules not compressed; stylus 2–3 cells wide at base; leaf cells commonly with intermediate thickenings; perianths without supplementary ridges, smooth or nearly so. 1. *F. eboracensis*
      4. Lobules compressed toward the lower, open end; stylus 3–5 cells wide at base; leaf cells with occasional intermediate thickenings; perianths with supplementary ridges (in addition to 1 postical and 2 antical keels), tuberculate. 2. *F. brittoniae*

1. *Frullania eboracensis* Gottsche [figs. 93, 94c] is relatively slender (with leafy stems and branches up to 1 mm broad) and dull-green or, when exposed to sunlight, brown or red-brown. The dorsal lobes are cordate at base. The lobules are saccate and helmet-shaped (except near female inflorescences), and the filiform stylus is only 2–3 cells wide at base. The leaf cells have distinct trigones and also intermediate thickenings. The underleaves are entire or occasionally obscurely unidentate at the sides. The plants are dioicous. The male spike has bracts in many pairs. The perianths, ± compressed and abruptly narrowed to a short beak, have 3 distinct keels but no supplementary ridges; the keels are smooth or nearly so. The spores, shed in May and June, sometimes germinate within the capsule to form large, multicellular, spherical bodies lacking a spore coat.

The name of the species, meaning York inhabiting, actually refers to New York, where the original collection was made.

This common *Frullania* grows on the bark of trees in moist woods, especially in swamps, in Clare, Eaton, Genesee, Gratiot, Huron, Ingham, Ionia, Isabella, Jackson, Kalamazoo, Lapeer, Livingston, Mecosta, Midland, Montcalm, Muskegon, Newaygo, Oakland, Saginaw, Sanilac, St. Clair, Van Buren, Washtenaw, and Wayne Counties.

All Frullanias look much alike. The only one at all common in Michigan is *F. eboracensis*. It is easily identified by helmet-shaped lobules, leaf cells with conspicuous corner thickenings and usually intermediate thickenings as well, and smooth perianths that are dorsiventrally flattened and 3-keeled. The male inflorescences are long and narrow, resembling a flattened corn cob. (Near the female inflorescence the lobules tend to be explanate rather than saccate.)

Forms with deciduous leaves are common. They sometimes bear globose gemmae at leaf margins, and these sometimes proliferate leaf primordia while still attached. Such forms should not be confused with *Frullania bolanderi* Aust. [fig. 94d], a black liverwort common on the trunks of roadside sugar maples especially near the shores of Lake Michigan farther north. That species produces an abundance of curved-erect branches made comblike by deciduous leaves but persistent, wide-spreading underleaves. The lobules may be one-half to three-fourths the size of the lobes.

2. *Frullania brittoniae* Evans [fig. 94a] is relatively large (the leafy stems and branches up to about 1.8 mm broad) and red-brown or greenish. The dorsal lobe is cordate at base. The uniformly saccate lobules are flattened toward their lower, open ends but not broader than long. The stylus is 3–5 cells wide at base and subulate. The leaf cells have conspicuous trigones but only occasional intermediate thickenings. The underleaves are broader than long and irregularly dentate or crenate-dentate on the sides. The plants are dioicous. The male spikes have many pairs of bracts. The perianths are strongly obovoid, abruptly narrowed to a slender beak and have supplementary ridges in addition to 3 keels; the upper surface and especially the keels are strongly tuberculate. Spores in May and June.

The species bears the name of Elizabeth Gertrude Britton, a shrewdly competent and internationally respected bryologist. It was her husband, Nathaniel Lord Britton, who founded the New York Botanical Garden.

The plants have been found on trees in Eaton and Washtenaw Counties. The species was reported by Irma Schnooberger (1940) from maple and elm trunks in Gratiot and Montcalm Counties (but no such specimens are to be found in her herbarium at the University of Michigan).

In contrast to *F. eboracensis*, *F. brittoniae* has lobules compressed toward their open ends, underleaves broader than long and irregularly toothed at the sides, and perianths tuberculate above, provided with supplementary ridges, and ending in a more slender beak. *Frullania inflata* is similar to *F. brittoniae* in that some lobules are explanate and the saccate ones compressed, but its underleaves are essentially entire-margined.

3. *Frullania inflata* Gottsche [fig. 94b] is relatively slender (leafy stems and branches up to 1.2 mm broad) and usually green but also brownish green to red-brown. The dorsal lobe of the leaf is not cordate at base. The lobules are mostly saccate and compressed toward the lower, open end, but many of them are explanate, small, and lingulate. The leaf cells have small and inconspicuous trigones and, infrequently, intermediate thickenings. The granulose oil bodies are relatively numerous (4–16 per cell) and elliptic to nearly linear. The underleaves are not much broader than the stems and

entire or nearly so at the sides. The plants are monoicous. The male branch, situated near the involucre, is a short spike consisting of about 2 pairs of bracts. The perianths are abruptly narrowed to a short, broad beak and often have supplementary ridges; the keels and ridges are smooth or nearly so.

The name of the species was chosen because of a perianth presumed to be flattened when young but inflated with age.

Specimens from Gratiot and Ingham Counties were taken from the base of oak trees. The species was also reported by Steere (1942) from the trunk of a red maple in Livingston County and by Schnooberger (1940) from the base of a poplar in Ionia County. Those specimens, named by Alexander W. Evans, are not to be found in the herbarium of the University of Michigan where both Steere's and Schnooberger's Michigan collections are deposited.

Frullania inflata strongly resembles *F. eboracensis* except for autoicous inflorescences, leaf cells usually without intermediate thickenings, perianths with numerous plications, and short, rounded male inflorescences. Explanate lobules are frequent, but many stems fail to produce them. (According to McGregor, 1955, explanate lobes are nearly always present in young growth and also in plants of unusually moist habitats.) The inflated lobules are somewhat flattened toward the open end (as in *F. brittoniae*).

Frullania plana Sull. was mentioned by Steere in his *Liverworts of Southern Michigan,* as known from Michigan from a single collection. No locality was given, and no specimen has been found. The species probably does not occur in the Great Lakes region. Because Steere confirmed the determinations of many of Miss Schnooberger's collections, it seems likely that the report was a slip of the pen for *F. inflata* based on her 1940 report from Ionia County. (*Frullania plana* has tiny, saccate lobules, underleaves cordate at base, and leaf cells commonly with intermediate thickenings.)

4. *Frullania riparia* Hampe ex Lehm. [fig. 94e] is relatively robust (leafy stems and branches up to 2 mm broad) and green or brownish green. The leaf lobes are cordate at the dorsal base. The lobules are usually explanate and lingulate (long and narrow as compared with those of *F. inflata*); the saccate lobules are wider than long but not particularly compressed. (On some plants all or nearly all the lobules

are explanate, but some plants have nearly all of them saccate.) The leaf cells have large trigones and occasional intermediate thickenings. The granulose oil bodies are 4–9 per cell and ellipsoidal. The underleaves are 3–4 times as wide as the stems, and entire or nearly so at the sides. The plants are dioicous, but perianths are unknown.

The name of the species refers, inappropriately, to banks of streams.

Specimens (from the bark of trees) are known from Gratiot, Kalamazoo, Muskegon, and Washtenaw Counties. Elsewhere in the Upper Midwest *F. riparia* seems to be limited to a rock substrate.

Both *F. riparia* and *F. inflata* have at least some explanate lobules, but smaller lobules, relatively large underleaves, and leaf cells virtually lacking intermediate thickenings give distinction to *F. inflata*, as does the narrow, flattened mouth of saccate lobules.

5. *Frullania asagrayana* Mont. [fig. 94f], also known as *F. tamarisci* subsp. *asagrayana* (Mont.) Hatt., is relatively large (with leafy stems and branches as much as 1.5 mm wide), somewhat spreading from the substrate, and shiny pinkish to red. The ventral lobes of leaves are saccate, and the dorsal lobes have a median streak of short, red cells in 1–2 rows extending more than halfway up the leaf. The rather large stylus is disc-shaped.

The original collection from Grandfather Mountain, North Carolina, was sent to Montagne for naming by Asa Gray.

This species is occasional on white cedar trunks in Northern Lower Michigan, at least as far south as Clare County (which lies partly within our range) and Iosco County (which lies slightly to the north).

Frullania asagrayana is sometimes considered a mere subspecies of the wide-ranging *F. tamarisci*, but I am loath to downgrade a species named for the first professor of botany at the University of Michigan!

L<small>EJEUNEACEAE</small>

26. Lejeunea Lib.

The plants are very small and yellow-green, with sparse branching and complicate-bilobed, entire leaves incubously arranged. The ven-

tral lobule is inflated and rolled toward the larger, rounded, somewhat concave dorsal lobe. The leaf cells are smooth. The underleaves are relatively large, rounded, and bilobed. The involucral
bracts are much like the leaves, and the perianths are oval to pear-
shaped and distinctly 5-angled toward a broad apex that is abruptly
narrowed to a minute beak. The capsules are spherical.

The generic name does honor to a Belgian physician-botanist,
Alexandre Louis Simon Lejeune.

Lejeunea cavifolia (Ehrh.) Lindb. [fig. 95] is monoicous (with male
bracts on a separate branch from the female inflorescence). The leaf
cells have 30–70 minute, simple oil bodies that are commonly long-
persistent. The underleaves are much larger than the lobules and
nearly contiguous. Spores in May and June.

The species' name appears to refer to hollowed out ventral leaf
lobes.

In northern Michigan *Lejeunea* grows on limestone rocks and
also on the trunks of white cedars, often in association with
Cololejeunea and *Radula*. It has been found at Grand Ledge, in Eaton County, on ledges of calcareous sandstone.

27. **Cololejeunea** (Spruce) Schiffn.

These minute, yellow-green plants have cortical cells of the stem
hyaline and thin-walled, in 1 layer, and leaves incubously overlapped and complicate-bilobed. The dorsal lobe is oblong or oblong-
ovate and broadly pointed; its firm, but rather thin-walled cells are
coarsely unipapillose on the dorsal surface. The ventral lobe, or lobule, is 1/2 as large as the dorsal lobe, or less. It is usually inrolled,
and its cells are thin-walled and smooth. A linear or sometimes
narrowly leaflike stylus is present at the junction of dorsal lobe and
stem. Underleaves are lacking. Rhizoids are attached to the stem
sparsely in clusters. Discoid, multicellular gemmae are common on
the ventral surface of the dorsal lobe. The male bracts are grouped in
loose terminal spikes; they are similar to leaves but somewhat more
inflated. The female bracts also resemble leaves, but they are less
spreading. The perianths are oval to pear-shaped, broad at the apex,

abruptly narrowed to a short-tubular mouth, distinctly 5-keeled, and strongly papillose. The capsules are spherical.

The name, meaning a mutilated *Lejeunea*, refers to the absence of underleaves.

Cololejeunea biddlecomiae (Aust.) Evans [fig. 96] has scarcely any rhizoids. The lobule is inrolled except at its upper end where there is a stout, 2-celled tooth. The stylus is linear and 5–8 cells long. The oil bodies are small, nearly smooth, and 7–12 per cell. The male spikes are at the ends of branches separated from the female inflorescences formed on very short branches.

The species was named for Hannah Biddlecome, of Akron and Columbus, Ohio, who collected it in Florida.

The species is rare and also easily overlooked because of its small size. It has much the yellow-green color of *Radula* and *Lejeunea* and grows in similar habitat niches. The plants, even smaller than *Lejeunea*, have been found on shaded rocks of wooded ravines and also on *Thuja* trunks in Lenawee, Washtenaw, and (according to Steere) Oakland Counties. (In Indiana, it is often found on the bases of sycamores in river bottom habitats.)

RADULACEAE

28. **Radula** Dum.

The plants, of medium size, grow in flat, yellow-green mats. The stems are loosely and irregularly pinnate, and the leaves are complicate-bilobed and incubously overlapped. The much larger dorsal lobe is rounded, and the ventral lobe is flat and subquadrate. Rhizoids are produced not on stems but on ventral lobes of leaves. Underleaves are none. The perianth is rectangular, dorsiventrally flattened, and square-tipped. The setae are short and the spores multicellular owing to precocious germination.

The generic name makes reference to scraperlike perianths. (The scraping tongue of a snail is called a radula.) Actually, the shape is much like that of a putty knife.

Radula complanata (L.) Dum. [fig. 97] is a very flat, yellow-green liverwort. Multicellular, discoid gemmae are commonly proliferated at the leaf margins that become ragged as a consequence. The oil bodies, 1–3 per cell (usually only 1), are very large (clearly visible with a scanning lens of a compound microscope), brown, and granulose-roughened. The plants have swollen male bracts directly below the female inflorescence, and perianths are abundantly produced. Spores in May to July.

The name of the species refers to leaves flattened together.

Radula, common on the leaning trunks of white cedar in northern Michigan, has been found in southern Michigan, often on that same substrate, in Clare, Eaton, Genesee, Gratiot, Huron, Ingham, Ionia, Isabella, Lapeer, Mecosta, Montcalm, Newaygo, Oakland, Van Buren, and Washtenaw Counties.

Class Anthocerotopsida (Hornworts)

The round, green rosettes are anchored on wet soil by 1-celled rhizoids. The thallus consists of uniform cells in a solid tissue except for antheridial and slime cavities. All the hornworts have ventral cavities that contain blue-green algae of the genus *Nostoc* [fig. 98c]. Some have, in addition, cavities scattered in the thallus tissue. The thin-walled cells, lacking trigones, contain one (or in some genera more than one) large, platelike chloroplast with a central group of starch-accumulating pyrenoid bodies. Slime papillae and oil bodies are lacking. Both antheridia and archegonia are internal. The antheridia occupy chambers, singly or in groups [fig. 98b]. The archegonia are embedded in thallus tissue [fig. 98a] but have necks reaching the upper surface; the neck cells are in 6 rows. The sporophytes consist of a massive foot and an elongate, usually long-cylindric (hornlike) capsule that dehisces from the apex downward into 2 valves. The capsule, sheathed at base by a collarlike involucre [fig. 99a], generally continues growth and spore production throughout the growing season owing to activities of a basal meristem [fig. 99a]. (The involucre completely encloses the young sporophyte but becomes collarlike because of rupture by the growing capsule.) The capsule wall is an essentially solid tissue of several cell layers, generally with stomata enclosed by 2 guard cells in the epidermal layer. The guard cells and inner layers of the wall have chloroplasts. The sporogenous tissue (derived from the amphithecium, fig. 102c–g) surrounds and overarches a slender columella of sterile cells [figs. 99b–d, 102g]. The large, tetrahedral spores are enmeshed in pseudelaters of 1-several cells that are only rarely marked by spiral thickenings. The spores produce only one gametophyte, and the protonematal stage is virtually lacking.

Hornworts are so different from mosses and liverworts that it is logical to separate them off as a division just as ferns have been isolated from the "fern allies." Schuster (1984) considered them to be early invaders of land, not at all related to other land plants and probably derived from a different group of algae. Crandall-Stotler (1984) considered mosses, liverworts, and hornworts as evolutionary analogues that have "adapted to the similar selection pressures imposed by a gametophyte-dominated, heteromorphic, archegoniate life cycle." She considered them "at most, remotely related, independently established, parallel evolving land plant groups." Her arguments and Schuster's too are convincing. It must be granted that little more than the relative dominance of gametophyte over sporophyte and their attachment to one another, as well as similarities in archegonia, antheridia, and sperms, justify the traditional view of an inclusive Bryophyta. However, the few similarities that hornworts have to mosses and liverworts are important ones, and the possibility of phylogenetic ties, however remote, is not easy to shrug off. (We might argue that *Sphagnum* is only slightly related to other mosses, and certainly *Bryum* is a far cry from *Marchantia*. Likewise, we might question whether man and manatees and monotremes really belong in the same class, Mammalia, or even in the same phylum, Chordata.)

Phaeoceros, often segregated from *Anthoceros*, has only ventral slime cavities, antheridial jackets consisting of many small cells, and yellow spores, whereas *Anthoceros* has been used in a more restricted sense for species with slime cavities scattered throughout the thallus, antheridial jackets with four tiers of elongate cells, and blackish spores. Both *Anthoceros* and *Phaeoceros* have long, slender, erect capsules that continue growing from the base and pairs of chloroplasts in the cells of the wall internal to the epidermis and in the paired guard cells enclosing stomata. There is a central columella, and the pseudelaters, consisting of several cells, generally lack spiral thickenings. The capsules of *Notothylas* are, by contrast, short and horizontal, with no basal meristem, essentially no columella, and no stomata. Also, chloroplasts are lacking in the capsule wall, and the pseudelaters separate into irregularly rounded, 1-celled structures with poorly marked spiral thickenings.

Gametophytic growth results from divisions of apical cells situated in notches at the thallus margin, though never from a tetrahedral cell with three cutting faces as in so many of the liverworts. In *Anthoceros* and *Notothylas*, at least, a wedge-shaped apical initial with four cutting faces cuts off segments in a spiral sequence.

Branching takes place when an apical cell repeatedly divides to form a transverse row of cells with two adjacent cells or cells at either side of the row taking over the function of cell initials. In either case, subsequent divisions result in a small lobe of tissue separating the new apical cells. In this way, equally forked, or dichotomous branching [fig. 100] results. Repeated branching around the thallus margin forms a rosette of a rounded, though lobed outline.

Unlike liverworts, hornworts produce antheridia internally. Antheridial production [fig. 101] begins with the division of an epidermal cell into two. A slime-filled cavity develops between the upper and the lower cells and eventually becomes an enlarged antheridial chamber [figs. 98b, 102b]. One to many antheridia may originate from a single cell initial (derived from the lower cell of the primary division) on the floor of this chamber, and secondary antheridia may also proliferate from the base of mature ones. Divisions of the upper cell form a roof over the antheridial chamber. (The roof eventually breaks down and exposes mature antheridia in conspicuous orange groups. The color is imparted by the antheridial jackets containing green plastids that turn orange as chlorophyll breaks down to expose carotene and xanthophyll.

Archegonial initials are epidermal in origin in hornworts (as in all other archegoniate plants), but division patterns result in their internal growth [figs. 101b, 102a]. Initiation begins by a transverse division of an epidermal cell. The outer cell undergoes vertical divisions to form a central cell enclosed by three outer ones. The central cell, by further transverse divisions, forms a vertical file of four or five neck canal cells and, at the base, a larger egg cell. Further divisions of cells surrounding the egg and neck canal produce an ill-defined jacket consisting of six vertical rows of cells, as in the Takakiales, Marchantiales, Sphaerocarpales, and Monocleales, in contrast to four in the Calobryales and five in the Metzgeriales and Jungermanniales.

The first division of the zygote is vertical rather than transverse (as in liverworts and mosses), apparently setting up from the beginning a bilateral symmetry carried through to the differentiation of two sutures along which dehiscence of the capsule takes place. A cell quadrant is soon set up, and its divisions form the endothecium that becomes a columella and the amphithecium that gives rise to the cell wall and the sporogenous tissue [fig. 102c–g]. The sporogenous tissue both surrounds and overarches the columella.

In Michigan, hornworts are found especially in the fall of the year on the soil of stubble fields, and uncommonly at that. In these habitats they are free of competition from larger plants, especially weeds, in both fall and spring. However, in Kansas, according to McGregor (1955), they grow in a variety of pioneering habitats, especially in wet prairies, over longer periods of time (*Anthoceros crispulus*, July 18 to October 2; *A. punctatus*, April 4 to November 24; *A. laevis*, June 18 to October 26; and *Notothylas orbicularis*, May 1 to November 24). Very likely longer seasons are possible in Michigan too, although summer-wet habitats are not common.

1. Capsules erect, long-cylindric, sheathed by the perigynium only at the base; cells of the capsule wall very long, only moderately thick-walled; stomata present.   1. *Anthoceros*
1. Capsules horizontal, shortly spindle-shaped, and almost completely sheathed by the perigynium; cells of the capsule wall subquadrate to short-rectangular and thick-walled; stomata none.   2. *Notothylas*

ANTHOCEROTACEAE

1. **Anthoceros** L.

The rounded thallus may be broadly short-lobed and flat-margined or deeply dissected and crisped at the margins. The lower epidermis has temporary slits near the growing tips of thallus lobes that allow the entrance of *Nostoc* into slime cavities. The capsules, sheathed at base by a tubular involucre, or perigynium, are erect, tall-cylindric, and green, with a solid wall of several layers of cells and a slender columella. The epidermal cells are long-rectangular, with moderately thickened walls and no chloroplasts, except that the paired guard cells enclosing the stomata contain chloroplasts and accumulate starch. The inner cell layers have 2 chloroplasts per cell. The capsule continues growth at the base while dehiscing from near the tip downward. After dehiscence the valves of the capsule may become irregularly twisted. The sporogenous tissue surrounds a slender columella. The pseudelaters are 1–5 cells long, variously curved or bent, and generally lacking in spiral wall thickenings.

The generic name refers to the long capsules as hornlike "flowers." Those species with yellow spores, non-tiered antheridial jackets [as in fig. 11b], and only ventral slime cavities are often segregated as *Phaeoceros*. *Aspiromitus* has been used, incorrectly, for species with black spores, tiered antheridial jackets, and thalli with scattered cavities that have traditionally been placed in *Anthoceros*. *Anthoceros* is used here in a sense inclusive of *Phaeoceros*.

All hornworts have on the undersurface small clefts opening into alga-containing slime cavities. The clefts can be seen only in the new growth near the lobe tips, as they close up after the algae gain entrance. *Anthoceros laevis* has only ventral slime cavities, but *A. punctatus* has, in addition, cavities scattered throughout the thallus formed by internal splitting followed by cell divisions. With age, such cavities often become air chambers. Only the ventral cavities contain colonies of *Nostoc*. (It is assumed that the algae fix atmospheric nitrogen, as they do when living free of the hornwort.)

1. Thalli pale-green, deeply dissected and wavy-margined, with slime cavities numerous and scattered throughout; capsules blackish or dark-brown; spores blackish (as seen in mass), densely spiculose.   1. *A. punctatus*
1. Thalli dark-green, with broad, flat lobes and slime cavities at the lower surface; capsules yellowish; spores yellow, finely papillose.   2. *A. laevis*

1. *Anthoceros punctatus* L. [fig. 103a–b] produces an abundance of long, slender capsules that resemble a clump of grass seedlings. The pale-green thalli are moderately divided and crisped at the margins and about 0.5–1.5 cm broad and 8–12(20) cells deep. The dorsal surface is often minutely warty and often bears a few lobules. Scattered in the thallus are many cavities. The plants are monoicous. The capsules come to be very long (up to about 5 cm), almost threadlike, and dark-brown or blackish. The spores are black (as seen in mass), 43–50 μm, and densely spiculose. Spores in August to November.

The plants have been found in Monroe, Washtenaw, and Wayne Counties on wet soil in fields harvested for corn and soy beans and also in winter wheat fields.

The specific epithet, meaning dotted, probably refers to a scattering of warty outgrowths (not lobules) sometimes seen on the dorsal surface. (Müller has explained the name on the basis of dark *Nostoc* colonies, which are, however, to be seen in all the species of *Anthoceros*.)

Anthoceros crispulus (Mont.) Douin has small, pale, yellow-green thalli that are more dissected and crisped (because of an abundance of marginal as well as surface lobes). It is said to have capsules 50–70 times longer than broad (in contrast to 30–40 times longer than broad in *A. punctatus*) and pseudelaters 3–5 cells long and as much as 210 μm in length (as opposed to those of *A. punctatus* that are often only 1–2 cells long and 60–100 μm in length). Though distinctive in appearance, *A. crispulus* may not be worth segregating from *A. punctatus*. It has been said that *A. crispulus* is no more than a response to unusually dry habitat conditions, but McGregor (1955) demonstrated that in Kansas *A. crispulus* occupies wetter habitats than those characterizing *A. punctatus*. *Anthoceros crispulus* has been reported (by Mayfield et al., 1983) from Monroe, Washtenaw, and Wayne Counties in the same kind of habitats as *A. punctatus*, but only two poor specimens (from Monroe and Wayne Counties) are represented in the herbarium of the University of Michigan, neither convincingly different from *A. punctatus*.

2. *Anthoceros laevis* L. [fig. 103c–h], also known as *Phaeoceros laevis* (L.) Prosk., has dark-green thalli 1–3 cm broad, with broad, flat lobes and slime cavities restricted to the lower surface. The plants are dioicous, and the male plants are somewhat smaller than the female. The mouth of the perigynium is commonly flared. The capsules seem to be yellowish because of spores showing through the wall. The spores are 42–50 μm and finely papillose. Spores from early June to early December.

The name of the species may refer to thalli not much dissected or crisped at the margins and therefore smooth or perhaps to an absence of dorsal lobules.

Plants have been found in Eaton, Jackson, Kalamazoo, Monroe, and Wayne Counties, on wet soil in harvested corn and soy bean fields and also in fields of winter wheat.

NOTOTHYLACEAE

2. Notothylas Sull.

The pale-green or yellowish thalli are flat and crenate-lobed at the margins. Dark patches representing scattered slime cavities are often present. The spindle-shaped capsules are horizontal, yellow or reddish brown, and largely enclosed in a cylindric perigynium. They dehisce by 2 valves and have no stomata, basal meristem, or clearly defined columella. The cells of the capsule wall are rectangular and thick-walled, and chloroplasts are lacking. The smooth, yellow spores measure 38–40 μm, and the pseudelaters are about the same size as the spores, 1-celled, and irregularly globose with very weak, slanted or more or less spiral markings.

The generic name refers to the perigynia as dorsal bags, and the specific epithet of our species refers to rounded thalli.

Notothylas orbicularis (Schwein.) Sull. [fig. 104], recognized by its narrow, horizontal perigynia and shortly emergent capsules, grows on heavy, wet soil in old fields, often at the corners of corn, soybean, and wheat fields, in ruts and tractor tracks, in pastures trodden by cattle, and on muddy footpaths. It is known from Kalamazoo, Lenawee, Monroe, Oakland, and Washtenaw Counties. Spores in August to November.

> An end is come, the end is come; it watcheth for thee; behold it is come.
>
> —Ezekiel

Glossary

I was never so bethumped with words.

—Shakespeare

acute sharply pointed at an angle less than 90°.

adnate grown together, fused.

adventitious originating out of the usual place.

air chamber a cavity in the spongy, green upper tissue of most of the Marchantiales; **air spaces** narrow, vertically oriented cavities in the green upper tissue of many species of *Riccia*.

amphithecium the outer layer of cells of a young sporangium. The first divisions of a young capsule produce a quadrant of four cells. Tangential divisions delimit an outer amphithecium and an inner endothecium. In liverworts, the spores arise in the endothecium; in hornworts, they originate in the amphithecium.

androecium a male inflorescence.

annular ringlike. The capsule walls commonly consist of cells with annular, semi-annular, or nodose thickenings.

androcyte a cell that becomes by further differentiation a sperm cell.

androgonial a small cell, referring to cells that divide to produce sperm cells.

antheridium a male sex organ, producing sperm cells.

antical in front, toward the observer and away from the substrate.

apical cell a cell initial, a cell at the growing point that divides repeatedly to form new tissues. Bryophytes grow by the activities of a single cell initial,

rather than by a meristematic tissue (except in the sporophyte of most hornworts); **apiculate** ending abruptly in a small point.

appendiculate having a subordinate part or attachment, such as a lobe on a ventral scale of the Marchantiales.

approximate close together.

archegonium a female sex organ, producing an egg; **archegoniates** cryptogamic plants producing eggs in multicellular archegonia, in other words, bryophytes, ferns, and fern allies.

archesporium the tissue of an embryonic capsule from which spores and elaters are differentiated.

areolate divided into small areas, often forming a network. The word is often used to refer to the cellular network of a leaf or thallus, also to the pattern of ridges ornamenting spores.

auriculate with a small, earlike lobe, or **auricle.**

autoicous monoicous but having antheridia on a branch separate from the stem or branch on which the archegonia are formed.

axil angle between leaf and stem.

bi- a prefix meaning 2; **bidentate** 2-toothed; **bifid** divided into 2 parts, forked; **biflagellate** having 2 whiplike cilia, or flagella; **bilobed** divided into 2 parts; **biseriate** in 2 rows; **bispiral** having double spiral thickenings; **bistratose** in 2 layers.

botryoid resembling a cluster of grapes.

bract a modified leaf associated with

male and female inflorescences; **bracteole** a modified underleaf of a female inflorescence. (Two bracts and often one bracteole make up an involucre.)

calciphile a plant of limy (calcareous, or marly) habitats.

calyptra a protective cover of a developing sporophyte derived from the base of the archegonium or, in liverworts, from the venter.

cambial activity cell divisions of a lateral meristematic tissue, or cambium, contributing to growth in diameter (not found in bryophytes).

capitate headlike.

capsule the sporangium, or spore-bearing portion of the diploid, asexual generation.

cell initial an apical cell that provides for growth owing to continued divisions.

centripetal proceeding toward the center, developing first toward the outside, later toward the inside (the opposite of **centrifugal**).

chloroplast a cellular inclusion that contains chlorophyll and carries on photosynthesis.

chromoplast a cellular inclusion containing yellow or orange pigments, a plastid low in chlorophyll and high in carotenoids.

cilium a hairlike structure; **ciliate** fringed with hairlike segments.

clavate club-shaped, larger toward the upper end. **Clavate** refers to a plane, **claviform** to a solid.

clone a group of individuals of identical genetic makeup resulting from vegetative propagation from a single individual.

coalesced fused together.

columella the central column of the capsule of a hornwort, surrounded by spores and pseudelaters.

complicate folded together; **complicate-bilobed** having 2 leaf lobes folded against one another.

compressed flattened; **dorsiventrally compressed** flattened at front and back; **frontally compressed** flattened at the front; **laterally compressed** flattened on both sides.

conidialike produced in a chain, as though pinched off in succession at the tip of a threadlike structure.

connate fused together, connected.

connivent approaching or converging at the tips, pincerlike.

cordate heart-shaped.

cortical referring to the outer part of a stem, or cortex.

crenate with close-set, rounded marginal teeth; **crenulate** finely crenate-toothed.

crisped strongly wavy, crinkled.

cruciate in the form of a cross.

cuticle the non-living covering over epidermal cells.

deciduous temporary, falling off.

decorticated lacking bark.

decurrent extending down the stem at 1 or both margins of a leaf.

decurved bent downward.

deflexed bent backward or downward.

dehiscent breaking or splitting open to release spores.

dentate toothed, with teeth spreading outward from the margin (as opposed to serrate, with teeth directed toward the apex); **denticulate** minutely dentate.

determinate growth growth that is limited, as in leafy liverworts where the apical cell is used up in producing archegonia.

dichotomous equally forked. True dichotomy (perhaps not found among liverworts) results from an apical cell that divides into two cells that also divide to form two new apical cells separated by a lobe of inactive cells. (See also **pseudodichotomy**.)

diploid having a double set of chromosomes.

dioicous having antheridia and archegonia on separate plants. (**Dioecious** is a comparable term applied to the sporophytic sexuality of higher plants.)

discoid flat and rounded, shaped like a disc.

divaricate forked, parted into 2 branches.

divergent spread apart.

dorsal upper; **dorsal furrow** or **groove** a channel on the upper surface of the thallus of the Marchantiales; **dorsal lobe** the upper division of a complicate-bilobed leaf.

dorsiventral flat, top-bottom in organization and structure.

egg the female gamete, produced in an archegonium.

elaters long, slender, hygroscopic, 1-celled structures with spiral thickenings mingled with spores of most liverworts, derived by mitotic divisions that differentiate them from meiotically derived spore mother cells; **elaterophore** a mass of cells to which some of the elaters are attached, either in a basal tuft or in tufts at the tips of valves of the capsule.

elliptic oval, somewhat longer than broad, with sides curved outward; **ellipsoid** a solid with the outline of an ellipse.

emarginate having a small notch or shallow indentation at the apex.

embryo a young sporophyte formed by the union of egg and sperm.

endo- a prefix meaning inner, inside; **endothecium** the inner cells of a young sporophyte of mosses and hornworts, see **amphithecium**.

endophytic living inside a cell.

entire lacking marginal teeth.

epidermis the outermost layer of cells.

erose irregularly worn away, eroded.

excentric off center.

exogenous formed externally.

explanate flattened out.

exserted protruding, projecting.

fascicle a bundle.

fertilization the union of sperm and egg, syngamy.

flagellum a slender, runnerlike branch with reduced leaves; **flagellate, flagelliform** slender, whiplike or runnerlike.

flask-shaped bottle-shaped, with a broad base contracted to a long, narrow neck.

foot the enlarged absorptive organ by which the sporophyte is anchored to the gametophyte.

frondose broad and flat, in reference to branching, resembling the compound leaf of a fern or a palm.

fruit sporophyte or capsule.

fusiform spindle-shaped, slender and tapered at both ends.

gamete sexual cell, egg or sperm; **gametophore** the gamete-bearing phase of the gametophyte, as opposed to the juvenile protonema or sporeling phase; **gametophyte** the haploid, sexual generation, producing gametes (eggs and sperm cells) in archegonia and antheridia. The gametophyte consists of the entire haploid generation from the spore onward and includes both juvenile and the gamete-bearing phases.

gemmae small bodies capable of reproducing a plant asexually; **gemmaecup** a rimmed depression in the thallus of *Marchantia* and *Lunularia* producing gemmae like eggs in a basket; **gemmiparous** bearing gemmae.

geniculate abruptly bent (like a knee).

glaucous whitish or with a whitish overcast.

globose, globular spherical.

granulate finely papillose, grainy in appearance.

guard cells cells surrounding a stoma, especially applied to the paired, kidney-shaped cells in an *Anthoceros* capsule wall.

haploid having a single set of chromosomes.

heteroicous monoicous, having sex organs variously located on the same plants or in other plants of the same population.

hyaline   colorless and transparent.

hygric   wet; **hygroscopic**   sensitive to moisture, moving in response to changes in humidity.

imbricate   overlapped, like shingles on a roof.

immersed   completely below the surface of an enveloping organ (as a calyptra immersed in a perianth).

incised   having a margin deeply, sharply, and irregularly notched or divided.

incubous   overlapped from the base of a stem toward the apex. (Each leaf overlaps the one next above.)

incurved   turned inward.

indeterminate   growth   unlimited growth.

inflorescence   the structure or structures enveloping sex organs, such as a perianth with its subtending involucre.

initial   a meristematic cell, often used in the sense of an apical cell.

insertion   attachment of a leaf to a stem or branch. (In leafy liverworts insertions may be transverse, incubous, or succubous.)

intermediate thickenings   nodulelike wall thickenings between the angles of a cell.

insertion   the line of attachment of a leaf.

intercalary   intermediate in position, somewhere below the apex.

involucre   a structure, usually tubular, surrounding or subtending archegonia or a calyptra containing a sporophyte in development; modified leaves of female inflorescences consisting of bracts and often bracteoles (comparable to leaves and underleaf).

keel   an abrupt or sharp fold.

lacerate   irregularly torn, deeply and irregularly cut.

lacinia   a narrow, incised segment of a leaf or involucre or a sharply cut, scalelike structure; **laciniate**   deeply and irregularly cut into fine divisions, as though slashed.

lacunose   marked by cavities, spongy.

lamella   a flat outgrowth from the surface.

lamina   a sheet of cells; thin wings on either side of the midrib of a thallose liverwort.

lanceolate   shaped like the head of a lance, broad at the base and tapered upward (similar to ovate but narrower).

lens-shaped   biconvex, lenticular.

ligulate   strap-shaped, long and narrow with parallel sides (sometimes used to mean tongue-shaped, or **lingulate**).

linear   long and narrow, with parallel sides, approaching a line in thickness.

lingulate   tongue-shaped, oblong below, broader toward the apex.

lobe   a major division of a leaf, larger than a tooth; **lobule**   the smaller of the lobes in species with complicate-bilobed leaves.

lumen   the cavity of a cell, enclosed by a wall.

lunar, lunate   shaped like a crescent-moon, crescent-shaped.

marsupium   a pouch developed after fertilization as a protection for the developing sporophyte.

median   middle.

meiosis   two successive cell divisions by which the chromosome number is halved and genetic segregation and recombination occur.

meristem   a tissue that continues cell division and growth (in *Anthoceros* at the base of the capsule).

merophyte   a block of cells cut off by the activity of the apical cell and giving origin to leaves, branches, and stem tissues (also called a segment).

mesic   moist, used in reference to environmental conditions or vegetational type (an intermediate condition contrasted with **hygric**, wet, and **xeric**, dry).

mitosis   ordinary cell division (as opposed to meiosis, or reduction division).

monoicous with antheridia and archegonia on the same plant, bisexual. (Bryologists prefer to use the words monoicous and dioicous for the haploid generation rather than **monoecious** and **dioecious,** terms better reserved for describing the sexuality of the sporophyte of higher plants.) See also **autoicous, heteroicous, paroicous,** and **synoicous.**

mucilage cavity a slime cavity, a space in the tissue of an *Anthoceros* thallus, often containing gelatinous masses of the blue-green alga *Nostoc;* **mucilage hair** a stalked slime papilla, a small and inconspicuous structure tipped with a gelatinous substance, often seen at leaf bases as a vestigial lobe, in place of an underleaf, or sometimes at the tips of leaf lobes.

multi- a prefix meaning many; **multicellular** many-celled; **multiseriate** in many rows; **multistratose** in many layers.

mycorrhiza a root-fungus symbiosis. The term is extended to rootless liverworts, such as *Aneura,* that have a similar symbiosis.

nodose with knobby thickenings. The cells of capsule walls often have small, nodose thickenings. See also **annular** and **semi-annular.**

ob- a prefix denoting the inverse condition; **obconic** with the shape of a cone but narrow at the base; **obcordate** heart-shaped but inverted, narrow at the base and broader above; **obovate** broader above, like an inverted egg.

oblong rectangular, longer than broad, with parallel sides.

obsolete very small and virtually lacking.

obtuse broadly pointed at an angle greater than 90°.

octant a ball of 8 cells formed early after fertilization in some liverworts.

oil bodies membrane-bound inclusions in cells of many liverworts, composed of terpenoid oils suspended in a proteinaceous matrix, of characteristic shapes, sizes, and colors (but generally colorless and glistening, rarely blue or brown); **oil cells** small cells of the Marchantiaceae containing a single, large oil body and no chloroplasts. The oil cells are generally scattered throughout the thallus, but they are often most obvious in the ventral scales, being colorless in contrast to the redness of other cells.

operculum lid or cap of a capsule, aiding in the dispersal of spores.

ostiole a small opening or hole.

oval elliptic, longer than broad, with sides curved outward.

ovate egg-shaped, broader at the base, gradually tapered above. (**Ovoid** refers to an egg-shaped solid rather than a plane.)

palmate lobed or branched in a radiating fashion, resembling a palm frond or fingers spreading from the palm of a hand.

papillose roughened by small protuberances of the cuticle.

paroicous monoicous, with antheridia situated near or below the archegonia on the same stem or branch (opposed to **autoicous,** with male and female sex organs on different branches of the same plant).

pellucid clear, translucent.

pendent hanging.

perianth a sheath surrounding the archegonia and developing sporophyte of leafy liverworts (representing an evolutionary fusion of leaves and underleaf).

perigynium a collarlike or tubular structure surrounding archegonia and a developing sporophyte formed by an upgrowth of thallus tissue.

pinnate branched in a featherlike fashion, with branches in a single plane on two sides of a stem.

plicate   folded lengthwise, pleated.

polygonal   with many angles and sides, usually ± isodiametric.

polyploid   having a chromosome number that is some multiple of the basic haploid number.

postical   behind, away from the observer, toward the substrate.

precocious germination   early germination as evidenced by segmentation of a spore into numerous cells before dispersal takes place.

primordium   the rudiment or beginning of an organ.

protonema   the juvenile phase of a gametophyte formed on germination of a spore (very much reduced in liverworts and hornworts and producing a single mature gametophyte); see also **sporeling**.

pseudelater   (more commonly **pseudoelater**) an angular, bent, or irregularly globose cell or series of cells mingled with the spores of hornworts.

pseudodichotomy   more or less equal forking resulting from an apical cell that cuts off segments that initiate a new apical cell, close to the original one.

pseudoperianth   a protective sheath surrounding one or more calyptrae and enclosed by an involucre in many of the Marchantiales.

punctate   dotted.

pyrenoid   a center for carbohydrate storage in chloroplasts of hornworts (and most green algae).

quadrant   a fourth part, often used in reference to the 4-celled stage of a young sporangium; **quadrate**   square.

receptacle   an expanded structure bearing archegonia or antheridia, often on a stalklike branch.

recurved   curved backward or downward.

reduction division   the halving of the chromosome number preceding spore formation.

reniform   kidney-shaped.

reticulate   forming a network.

retuse   notched at a broad apex.

revolute   rolled back (a more extreme condition than recurved).

rhizoid   a rootlike hair, in liverworts and hornworts 1-celled (in mosses many-celled); in most Marchantiales there are 2 kinds of rhizoids, smooth and internally pegged; **rhizoid furrow**   a rhizoid-bearing channel on 1 or both sides of the stalk of a receptacle of the Marchantiales.

rosette   a rounded, lobed thallus.

saccate   deeply concave, inflated, bag-like.

schizogenous   formed by cells splitting apart, used in reference to intercellular spaces such as air chambers of the Marchantiales formed by cells splitting apart.

sciophilous   living in the shade.

secondary growth   growth in diameter resulting from divisions of cambium cells (not found in bryophytes).

secund   turned to 1 side.

segment   division (or branch) of a thallus, or merophyte, a block of cells cut off by an apical cell; **segmentation**   cell division.

semi-annular   incompletely ringlike. Cells of capsule walls commonly have annular, semi-annular, or nodose thickenings.

serrate   with teeth directed toward the apex, saw-toothed; **serrulate**   minutely serrate.

sessile   without a stalk.

seta   stalk of a sporophyte.

shoot calyptra   a protective organ surrounding the developing sporophyte formed by a hollowed-out stem tissue crowned by a calyptra.

simple   unbranched or, in the case of oil bodies, unsegmented.

sinus   a depression separating 2 lobes.

slime cavity   a cavity filled with gelatinous material found in the thallus of hornworts; **slime papilla** a small and inconspicuous cell, often stalked, tipped with a gelatinous substance, often seen at leaf bases as a vestigial lobe, in place of an underleaf, or sometimes at the tips of a lobed leaf; see also **mucilage hair.**

sperm a male gamete, in bryophytes biflagellated; **spermatogenous tissue** cells which by their divisions give rise to sperm cells within the jacket of an antheridium.

spiculose finely and sharply papillose.

spindle-shaped fusiform, narrow and gradually tapered to both ends.

spinose coarsely toothed, the teeth long and spinelike.

spore a 1-celled reproductive body resulting from meiotic divisions in the sporangium, or capsule; **spore mother cell** a cell of the capsule that undergoes meiosis to produce 4 spores; **sporogenous tissue** the archesporium, or tissue from which spores and elaters differentiate; **sporophyte** the diploid, asexual, spore-bearing generation, attached to the gametophyte and, in liverworts, parasitic on it; **sporeling** the juvenile stage of a gametophyte, including the scant protonema and primordial tissues of an embryonic gametophyte.

squarrose spreading at a 90° angle.

stellate star-shaped.

sterile not fruiting, not bearing inflorescences or sporophytes.

stoloniform resembling a slender, runnerlike branch, or stolon.

stoma a pore (enclosed by 2 guard cells in hornworts, by concentric series of cells in many of the Marchantiales); **stomatal apparatus** or **structure** concentric rows of differentiated cells surrounding epidermal pores, or stomata, in many of the Marchantiales, sometimes called guard cells; if the cells of the stomatal apparatus are in 1 layer, the pores are **simple,** if in several layers, in a barrel-like arrange-

ment, they are **compound.** (Stomatal structure is a better designation than stomatal apparatus, not implying any function in opening or closing the stomata.)

strap-shaped long and narrow with parallel sides, ligulate.

striate with elongate grooves or scratches (much finer than plicate).

stylus a small appendage, usually linear, at the attachment of lobule to lobe in the complicate bilobed leaves of *Frullania* and *Cololejeunea.*

sub- a prefix meaning below or nearly, not quite, as in **subentire** (nearly entire), **suberect** (nearly erect), **subquadrate** (nearly square), **subspherical** (approaching the shape of a globe or sphere).

succubous obliquely inserted and, if crowded, overlapped from the apex downward (like shingles on a roof). (Each leaf overlaps the one next below.)

synoicous monoicous, having antheridia and archegonia mixed together in the same inflorescence.

terete smooth cylindric, rounded in section (not plicate or cornered).

tetrad a group of 4, used in reference to spores held together in groups of four as a result of their meiotic origins.

tetrahedral 4-sided, used in reference to spores shaped by tetrad groups in which they formed as a result of reduction division.

thallus a flat plant body, not differentiated into stems and leaves; **thalloid** flat, ribbbonlike or rosettelike.

transverse insertion the crosswise attachment of a leaf, as opposed to an oblique attachment, whether **incubous** or **succubous.**

tri- a prefix meaning 3; **trifid** divided into 3 parts; **trigone** a 3-angled thickening at the corners of cells; **trigonous** 3-angled or 3-sided (as in 3-cornered perianths).

truncate abruptly cut off or squared off at the apex.

tuberculate with small warts (coarser than papillae).

tumid swollen in appearance.

turgor plumpness resulting from internal pressure exerted on cell walls owing to the absorption of water.

underleaves a third row of leaves found on the lower, or ventral, surface of most leafy liverworts.

uni- a prefix meaning 1; **unipapillose** with 1 papilla per cell; **uniseriate** in 1 row; **unispiral** with a single spiral thickening; **unistratose** in 1 layer.

valve one of the longitudinal segments into which capsules split on dehiscence.

venter the expanded base of an archegonium containing the egg and becoming the calyptra; **ventral** lower; **ventral lobe** the lower division of a complicate-bilobed leaf; **ventral scale** a flat outgrowth among the rhizoids on the lower surface of most of the Marchantiales; **ventricose** bulging, bellylike.

verruculose finely and irregularly roughened (but in the literature on liverworts often used in the sense of papillose).

xeric dry (as opposed to **hygric,** wet, and **mesic,** moist).

zygote the first cell of the sporophytic generation, resulting from the fertilization of an egg by a sperm.

BIBLIOGRAPHY

Albert, D. A., S. R. Denton, and B. V. Barnes. 1986. Regional Landscape Ecosystems of Michigan. 32 pp. + map insert. School of Natural Resources, University of Michigan, Ann Arbor.

Bowers, F. D., and S. K. Freckmann. 1979. Atlas of Wisconsin bryophytes. Rept. Fauna Fl. Wis 16(1): 1–53. (Museum of Natural History, Stevens Point, Wis.)

Casares-Gil, A. 1919. Flora Ibérica. Briófitas (Primera Parte). Hepaticas. xx + 775 pp. Museo Nacional de Ciencias Naturales, Madrid.

Crandall-Stotler, B. 1984. Musci, hepatics and anthocerotes—an essay on analogues. pp. 1093–1129 in R. M. Schuster (ed.), New Manual of Bryology. Vol. 2. Hattori Botanical Laboratory, Nichinan.

Evans, A. W. 1902–23. Notes on New England Hepaticae I–XVII. Rhodora 4: 207–213 (1902); 6: 165–174, pl. 57, 185–191 (1904); 7: 52–58 (1905); 8: 34–45 (1906); 9: 56–60, 67–72, pl. 73 (1907); 10: 185–193 (1908); 11: 185–195 (1909); 12: 193–204 (1910); 14: 1–18 ,109–115 (1912); 16: 62–76 (1914); 17: 107–120 (1915); 18: 74–85, 103–120, pl. 120 (1916); 19: 263–272 (1917); 21: 150–169, pl. 126 (1919); 23: 281–284 (1921); 25: 74–83, 89–98 (1923).

Evans, A. W. 1910–23. Notes on North American Hepaticae. I–X. Bryol. 13: 33–36 (1910); 12: 84–88 (1911); 15: 54–63, pl. 2 (1912); 16: 49–55 (1913); 17: 87–92 (1914); 18: 81–91, pl. 1 (1915); 20: 17–26 (1917); 22: 54–73 (1919); 25: 25–33 (1922); 26: 55–67 (1923).

Flowers, S. 1961. The Hepaticae of Utah. Univ. of Utah Biol. Ser. 12(2): 1–89, pls. 1–18.

Hollensen, R. H. 1984. A checklist of the liverworts of Michigan. Mich. Bot. 23: 135–139.

Ireland, R. R., and G. Bellolio-Trucco. 1987. Illustrated Guide to Some Hornworts, Liverworts and Mosses of Eastern Canada. Syllogeus 62, 205 pp. National Museum of Canada, Ottawa.

Kauffman, C. H. 1915. A preliminary list of the bryophytes of Michigan. Rept. Mich. Acad. Sci. 17: 217–223.

Macvicar, S. M. 1926. The Student's Handbook of British Hepatics. Ed. 2. Sumfield, Eastbourne.

Mayfield, M. R., M. C. Cole, and W. H. Wagner, Jr. 1983. Ricciaceae in Michigan. Mich. Bot. 22: 145–150.

McGregor, R. L. 1952. Riccia rhenana Lorbeer in Kansas. Bryol. 55: 129–130.

McGregor, R. L. 1955. Taxonomy and ecology of Kansas Hepaticae. Univ. Kansas Sci. Bull. 37, part 1, no. 3, pp. 35–141.

Miller, N. G. 1972. A bryophyte florule of an area near Grand Ledge, Eaton County, Michigan. Mich. Bot. 11: 3–12.

121

Müller, K. 1905–16. Die Lebermoose Europas. *In* L. Rabenhorst, Kryptogamen-Flora von Deutschland, Oesterreich und der Schweiz. Ed. 2, Vol. 6(1), pp. 1–871, 6(2), pp. 1–947. Kummer, Leipzig.

Müller, K. 1951–58. *Ibid.* Ed. 3. Vol. 6, pp. 1–1365. Geest & Portig, Leipzig.

Nichols, G. E., and W. C. Steere. 1936. Notes on Michigan bryophytes – III. Bryol. 39: 111–118.

Renzaglia, K. S. 1978. A comparative morphology and developmental anatomy of the Anthocerotophyta. Jour. Hattori Bot. Lab. 44: 31–90.

Schnooberger, I. 1940. Notes on bryophytes of central Michigan. Pap. Mich. Acad. 25(1939): 101–106.

Schofield, W. B. 1985. Introduction to Bryology. 431 pp. Macmillan, New York.

Schuster, R. M. 1949. The ecology and distribution of Hepaticae in central and western New York. Amer. Midl. Nat. 42(3): 513–712.

Schuster, R. M. 1953. Boreal Hepaticae: a manual of the liverworts of Minnesota and adjacent regions. Amer. Midl. Nat. 49(2): 257–684.

Schuster, R. M. 1966–81. Hepaticae and Anthocerotae of North America, East of the Hundredth Meridian. Vols. 1–4. Columbia University Press, New York. [Vol. 5 & 6 not yet published]

Schuster, R. M. 1983. Reproductive biology, dispersal mechanisms, and distribution patterns in Hepaticae and Anthocerotae. Sonderb. Naturw. Ver. Hamburg 7: 119–162.

Schuster, R. M. 1983–84. New Manual of Bryology. 2 vol. Hattori Botanical Laboratory, Nichinan.

Smith, A. J. E. 1990. The Liverworts of Britain and Ireland. 380 pp. Cambridge University Press.

Steere, W. C. 1934. Unreported or otherwise interesting bryophytes from Michigan. Bryol. 37: 57–62.

Steere, W. C. 1940. Liverworts of Southern Michigan. 97 pp. Cranbrook Institute of Science, Bloomfield Hills, Michigan.

Steere, W. C. 1942. Notes on Michigan bryophytes-IV. Bryol. 45: 153–172.

Stotler, R., and B. Crandall-Stotler. 1977. A checklist of the liverworts and hornworts of North America. Bryol. 80: 405–428.

Watson, E. V. 1981. British Mosses and Liverworts. Ed. 3. xviii + 519 pp. Cambridge University Press.

Watson, E. V. 1971. The Structure and Life of Bryophytes. 211 pp. Hutchinson, London.

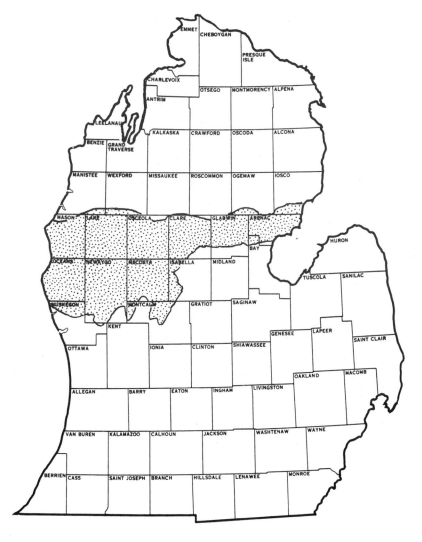

Fig. 1. Map of Michigan showing the 42 counties of Southern Lower Michigan, in other words, those counties lying south of the Tension Zone (shaded).

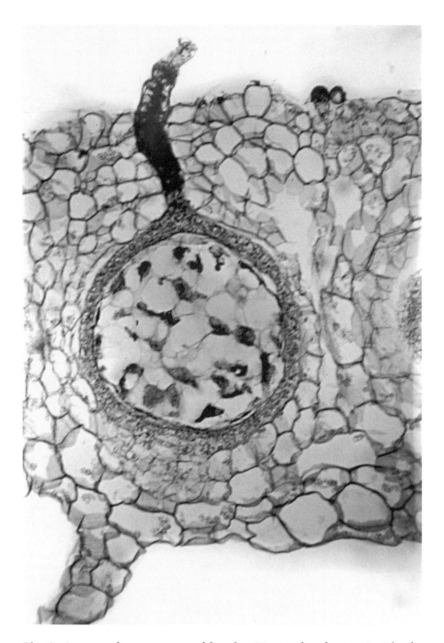

Fig. 2. A sporophyte, presumably of a *Riccia*, developing inside the enlarged venter of an archegonium, or calyptra. The neck of the archegonium projects beyond the upper surface of the thallus. The sporophyte in *Riccia* and *Ricciocarpos* is no more than a ball of spores enclosed by a 1-layered capsule wall.

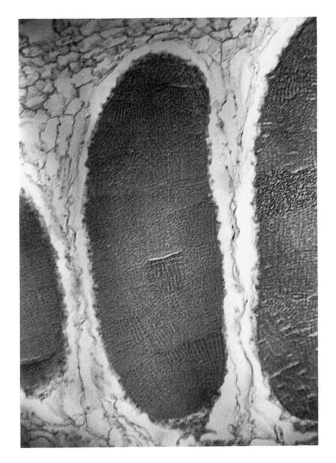

Fig. 3. Antheridium of *Conocephalum,* formed inside an air chamber at the dorsal surface of an antheridial receptacle. The cucumberlike antheridium has a basal stalk and a 1-layered jacket of cells enclosing innumerable androcytes, or cells that develop into biflagellate sperms, in sectored patches.

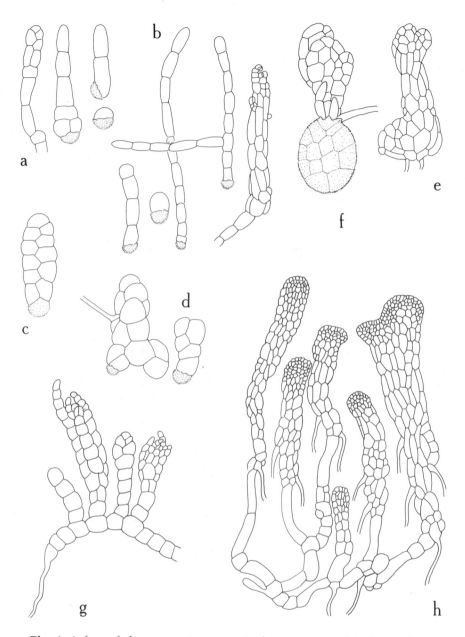

Fig. 4. A few of the many "patterns" of protonematal and sporeling development in liverworts. a. *Nardia sieboldii*, filamentous. b. *Cephalozia otaruensis*, filamentous; leaf primordia shown on axis at right. c. *Bazzania tridens*, cylindric. d. *Calypogeja tosana*. e. *Pellia epiphylla*. f. *Lepidolaena clavigera*, a cell mass developed after precocious germination within the spore wall. g. *Chiloscyphus polyanthos*, clonal growth. h. *Bucegia romanica*, clonal growth. (a–d, redrawn from Nehira, *Journal of the Hattori Botanical Laboratory*, 1974; e, from Fulford, *Phytomorphology*, 1975; f–h, from Chalaud in Verdoorn's *Manual of Bryology*, 1932, these in turn from earlier publications by Goebel and Teodorescu.)

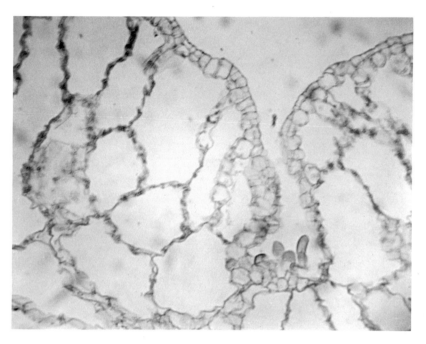

Fig. 5. The median groove of a *Ricciocarpos* near the growing point showing a cluster of mucilage hairs.

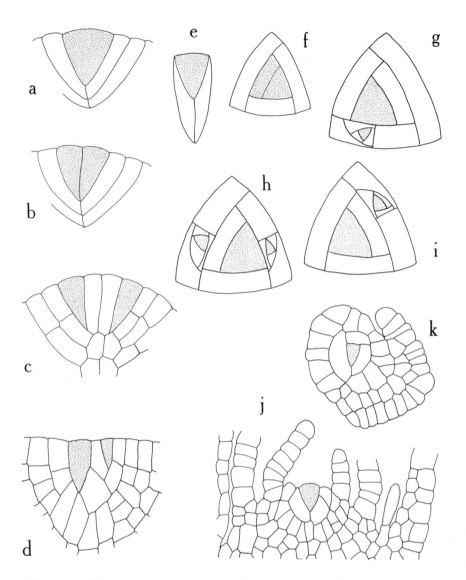

Fig. 6. a–c. Divisions of an apical cell leading to true dichotomy, as it may occur in the Marchantiales (and does in the hornworts). d. Divisions of an apical cell to form 2 cell initials of different ages leading to false dichotomy, as in some of the Metzgeriales. e. Tetrahedral apical cell with 3 cutting faces, typical of the Jungermanniales, as seen in its entirety. f. Segmentation of a tetrahedral apical cell as seen in cross section. g–i. Segmentation leading to the formation of leaves, underleaves, and "terminal" branches in the Jungermanniales (g, *Acromastigum* type; h, *Frullania* type; i, *Microlepidozia* type). j. Growth from a tetrahedral apical cell in *Scapania nemorea*, shown in longitudinal section; at right is a mucilage hair in the axil of a leaf primordium. k. Growth from the apical cell of *Scapania nemorea* in section. (a–c redrawn from Müller, *Lebermoose Europas*, g–i from Evans, *Annals of Botany*, 1912, j–k from Zehr, *Bryophytorum Bibliotheca*, 1980).

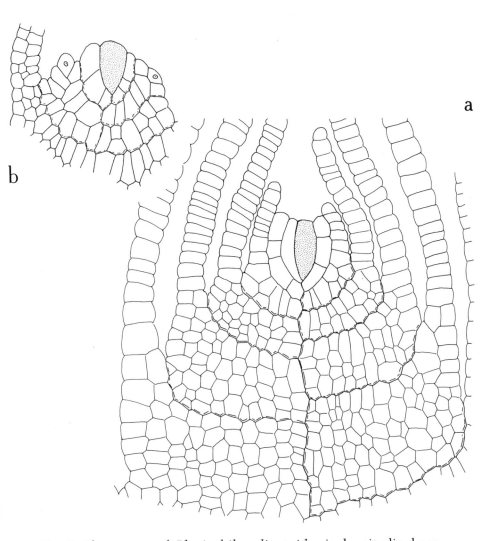

Fig. 7. Shoot apex of *Plagiochila adiantoides* in longitudinal sections. a. Apical cell, lateral merophytes, and bases of leaf rudiments. b. "Terminal" branch initial of the *Frullania* type in the second merophyte at the left and presumably an incipient branch initial at the right. (After Johnson, *Botanical Gazette*, 1929.)

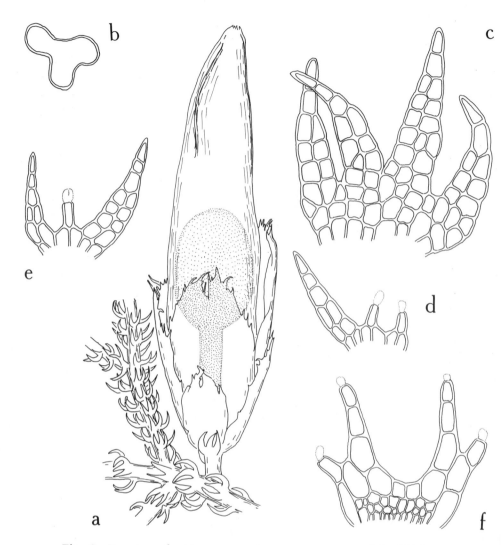

Fig. 8. *Kurzia sylvatica,* a species not represented in Michigan's flora, although a similar species, *K. setacea,* occurs in bogs of the North. a. Portion of a plant showing 4-lobed leaves and a perianth enclosing a nearly mature sporophyte and surrounded at base by bracts of the involucre. b. The upper portion of a perianth as seen in section. c. A leaf showing four lobes, each of them having cells 2- to 3-seriate at base (in contrast to *Blepharostoma,* which has lobes constructed of uniseriate cells); the leaf lobes are not tipped by slime papillae. d, e. Underleaves, showing slime papillae at the tips of rudimentary lobes. f. *Calypogeja sullivantii* underleaf, showing rhizoid-bearing cells at base and lobes and lateral teeth tipped with slime papillae. (Redrawn from Evans, *Rhodora,* 1902–23.)

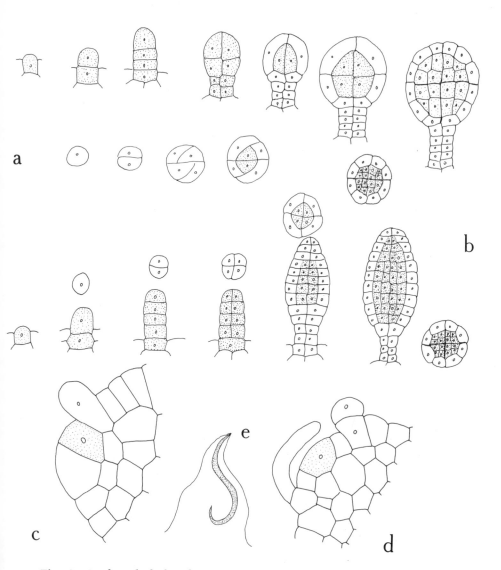

Fig. 9. Antheridial development. a. Metzgeriales and Jungermanniales. b. Marchantiales. c. Origin of antheridial initial close to the apical cell in *Fossombronia* (Metzgeriales). d. Antheridial initial, apical cell, and overarching slime papilla, in *Fossombronia*. (After Schuster, *New Manual of Bryology*, 1984.) e. Biflagellate sperm cell.

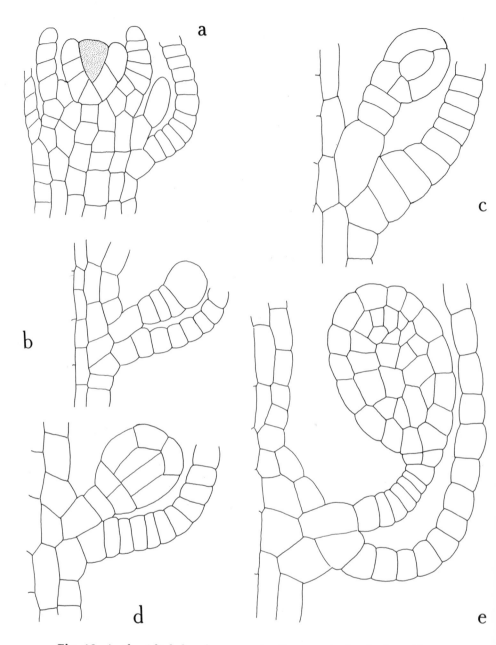

Fig. 10. Antheridial development in *Bryopteris fruticulosa* (Junger-manniales), as seen in longitudinal sections. a. Shoot apex, showing antheridial initial in the axil of a modified leaf, or bract. b. Differentiation of the stalk from the body of the antheridium. c–e. Stages in the differentiation of jacket and spermatogenous tissue. (After Stotler & Crandall-Stotler, *Bryophytorum Bibliotheca*, 1974.)

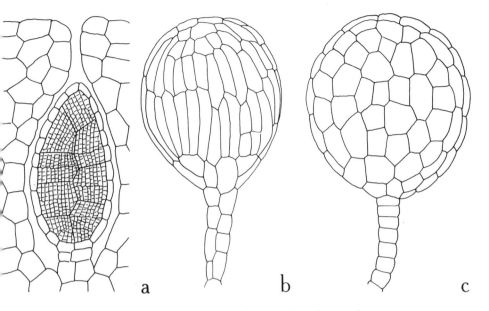

a   b   c

Fig. 11. Antheridial forms. a. A sunken antheridium of *Ricciocarpos* in section, showing the ellipsoidal form typifying the Marchantiales (redrawn from Smith et al., *A Textbook of General Botany*). b, c. Antheridia of the Jungermanniales, showing variations in the stalk and the number and shape of jacket cells. (The antheridium shown at *b* has 'tiered" jacket cells.) The antheridia of the Metzgeriales show an even greater variation. (Redrawn from Schuster, *New Manual of Bryology*, 1984.)

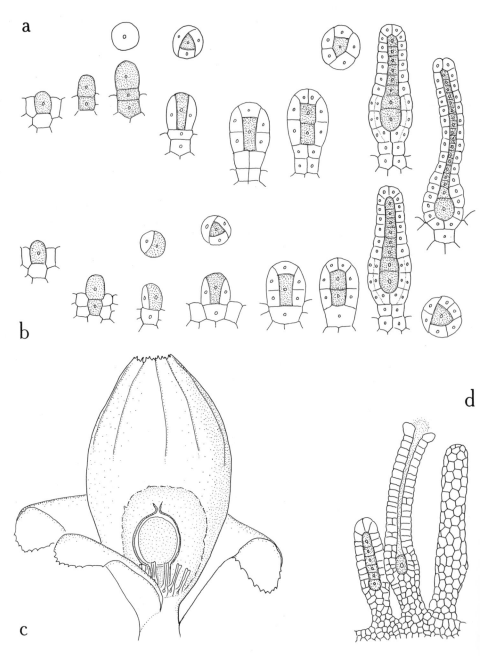

Fig. 12. Archegonial development. a. Metzgeriales and Jungermanniales. b. Marchantiales. (After Schuster, *New Manual of Bryology*,
1984.) c. Perianth, involucral bracts, archegonia, and young
sporophyte of *Diplophyllum*. d. Archegonia of *Plagiochila*. (After
Casares-Gil, *Flora Ibérica*, 1919.)

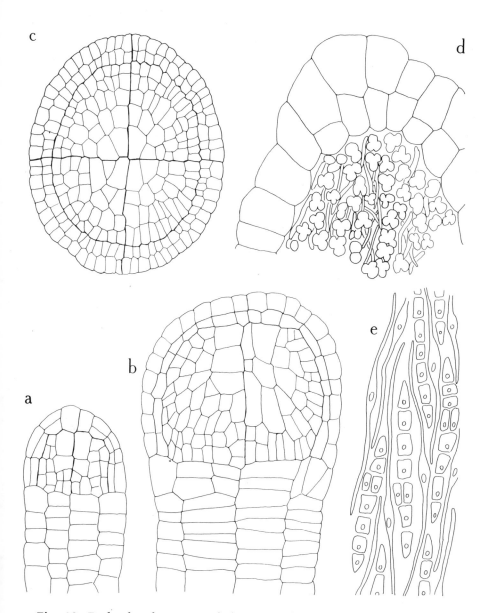

Fig. 13. Early development of the sporophyte in *Aneura ambros-ioides*. a–c. Divisions establishing the amphithecium from which the capsule wall develops and the endothecium from which sporogenous tissue develops. (Redrawn from Bower, *The Origin of a Land Flora*.) d–e. Differentiation of sporogenous tissue in *Monoclea forsteri* (Monocleales), showing features held in common with other liver-wort orders. d. Portion of capsule in section, showing young elaters mingled with presumably 4-lobed spore mother cells. e. Young elaters and linear series of spore mother cells secondarily formed by mitotic divisions. (Redrawn from Johnson, *Botanical Gazette*, 1894).

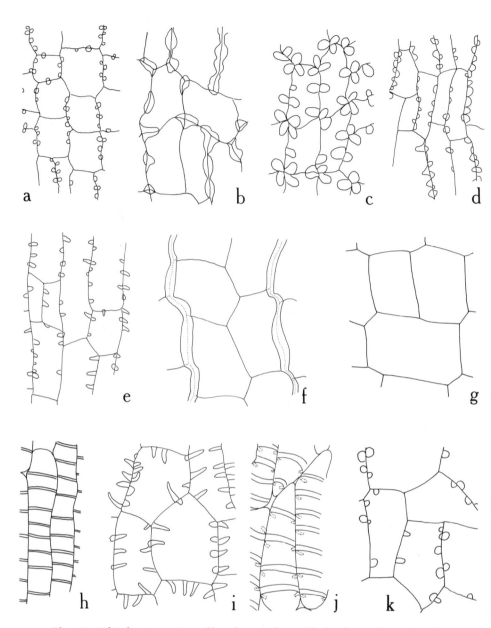

Fig. 14. Thickenings in cells of capsule walls (a–h, epidermal cells; i–k, inner cells). a. *Lophocolea heterophylla*. b. *Radula complanata*. c. *Frullania eboracensis*. d. *Lepidozia reptans*. e. *Cephalozia lunulifolia* . f. *Calypogeja muelleriana*. g. *Fossombronia cristula*. h. *Marchantia polymorpha*. i. *Jungermannia leiantha*. j. *Calypogeja muelleriana*. k. *Fossombronia cristula*.

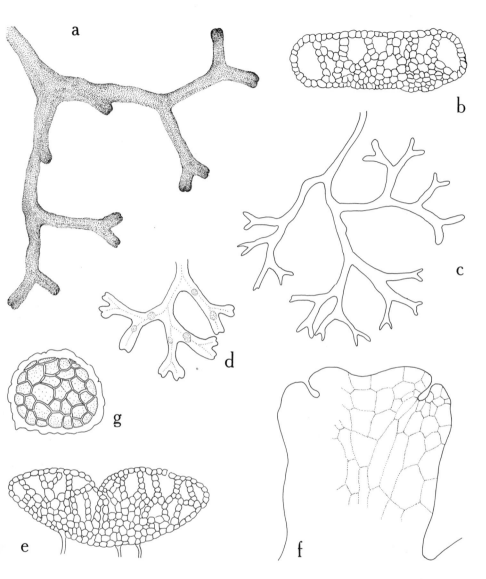

Fig. 15. a. *Riccia fluitans*, aquatic expression. b. Cross-section of *R. fluitans*, aquatic expression. c, d. *R. fluitans* and *R. canaliculata*, respectively, drawn to scale, *R. canaliculata* with sporangial swellings at the undersurface. e. Cross-section of *R. canaliculata*. f. Pattern of areolation in *R. canaliculata* (adapted from Schuster, *American Midland Naturalist*, 1953). g. Spore of *R. canaliculata*.

Fig. 16. Habits of species of *Riccia*. a. *R. frostii*. b. *R. hirta* (photo by Marie Cole). c. *R. beyrichiana* (photo by Wm. Brodowicz).

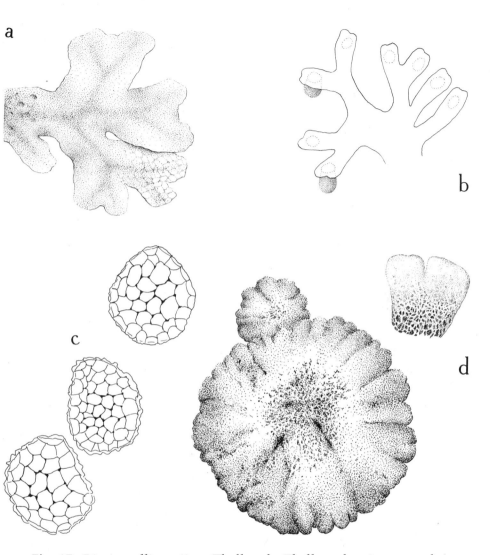

Fig. 17. *Riccia sullivantii.* a. Thallus. b. Thallus, showing ventral swellings where sporangia have formed. c. Spores. *Riccia cavernosa.* d. Thalli, with one lobe enlarged.

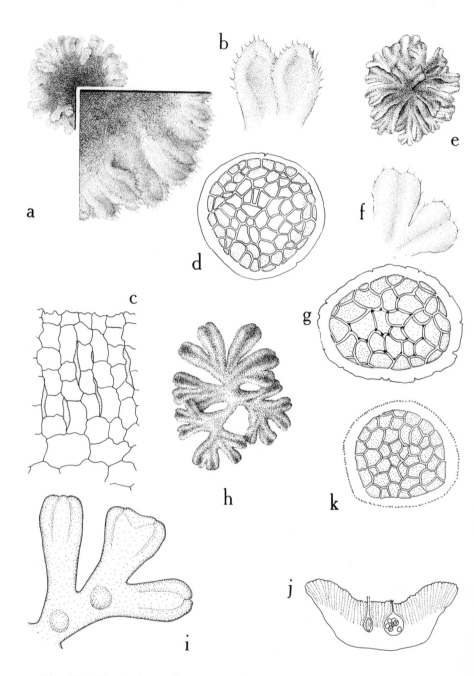

Fig. 18. *Riccia beyrichiana.* a. Thallus. b. Lobe tips, enlarged. c. Portion of thallus in cross-section showing vertical files of cells and narrow air channels among them; the terminal cell of each file becomes collapsed with age. d. Spore. *Riccia hirta.* e. Thallus. f. Lobe tips, enlarged. g. Spore. *Riccia bifurca.* h. Thallus. i. Thallus tips, enlarged, with sporophytes showing through upper surface. j. Thallus in section, showing an antheridium and an old sporophyte enclosed in the thallus tissue. k. Spore.

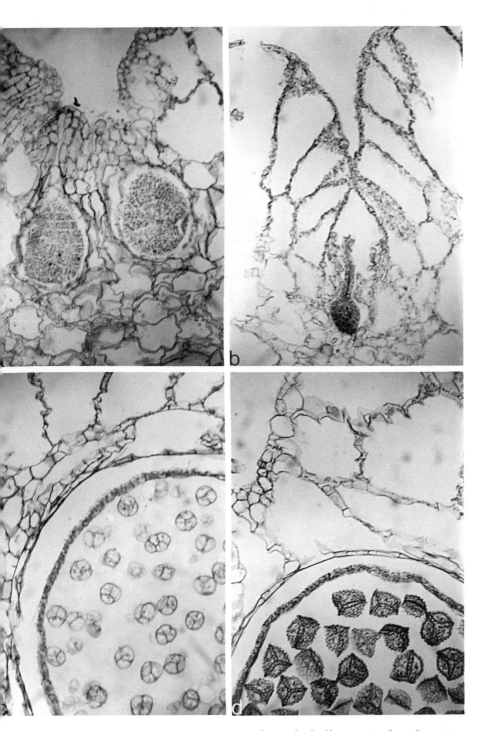

Fig. 19. *Ricciocarpos natans*, sections through thallus. a. Antheridia. b. Archegonium. c, d. Spore tetrads, immature, and mature spores showing triradiate scars developed inside the calyptra, the capsule wall having been absorbed, presumably as a nutrient for developing spores.

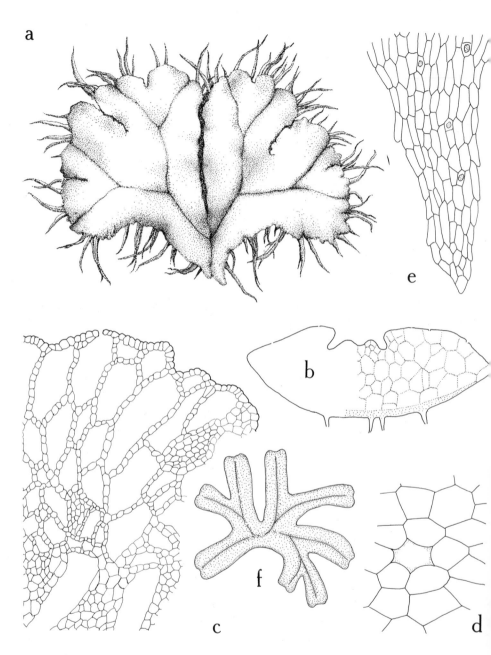

Fig. 20. *Ricciocarpos natans.* a. Thallus, floating form. b, c. Cross-sections of thallus. d. Upper epidermis, showing a pore. e. Portion of a ventral scale. f. Stranded form of thallus.

Fig. 21. *Ricciocarpos natans* floating on the shallow water of a pond.

Fig. 22. *Lunularia cruciata.* Thallus with gemmae only partially encircled by a lunate, or crescent-shaped rim (photo by Wm. Randolph Taylor).

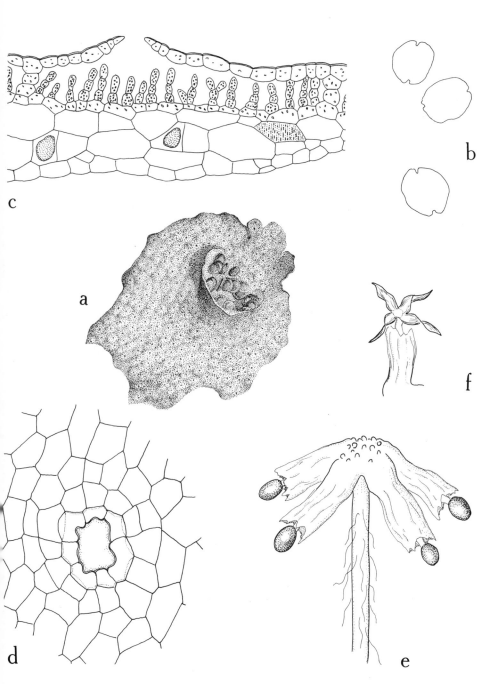

Fig. 23. *Lunularia cruciata*. a. Thallus, showing gemmae-cup. b. Gemmae. c. Cross-section of thallus, showing large oil bodies and faint markings on cell walls. d. Upper epidermal cells, with pore. e. Female receptacle and sporophytes. f. Capsule in dehiscence, showing some exsertion from the tubular involucre.

Fig. 24. *Conocephalum conicum.* a. Thallus showing seemingly dichotomous branches, coarsely sectored dorsal epidermis, and conspicuous, white stomatal structures. b. Conic female receptacle elevated on a stalklike branch (photo by A. J. Sharp). c. Discoid, sessile male receptacles (photo by Norton Miller).

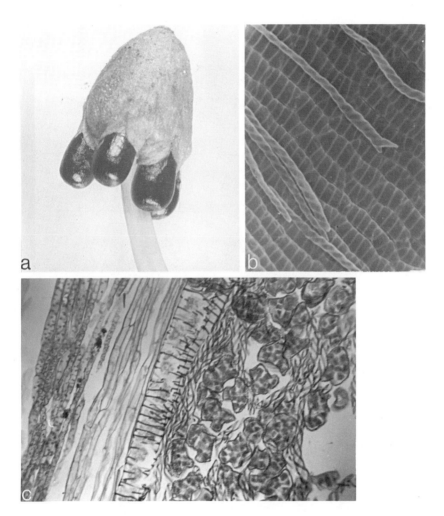

Fig. 25a–c. *Conocephalum conicum.* a. Shiny black capsules hanging from the undersurface of a female receptacle (photo by E. B. Mains). The spores are multicellular at maturity. b. Capsule wall and elaters. c. Portion of the capsule in section. The wall is 1-layered. Young spore tetrads lie among bispiral elaters. Pegged rhizoids, at left, belong to a rhizoid furrow extending from the stalk into the receptacle. (Photos b and c by Jane Taylor.)

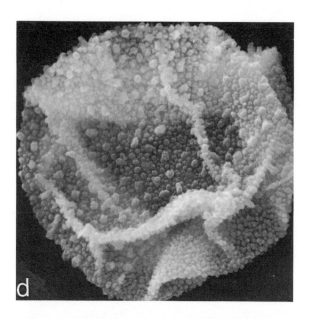

Fig. 25d–e. d. *Conocephalum conicum.* Spore, somewhat collapsed owing to treatment for SEM preparation and probably not showing the multicellular nature. (Photo by Jane Taylor.) e. Female receptacle of *Preissia quadrata,* showing pendent sporophytes and inflated pseudoperianths. The tiny, though distinct, stomatal structures are characteristic of the species. (Photo by Jeffrey Holcombe.)

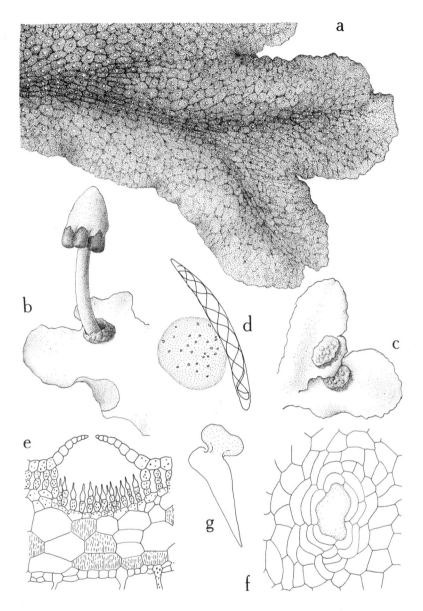

Fig. 26. *Conocephalum conicum.* a. Portion of thallus. b. Female inflorescence, showing sporophytes pendent from the undersurface of a conic receptacle. c. Discoid, sessile male receptacles. d. Spore and elater. e. Section through thallus, showing air chamber floored by green filaments, simple though conspicuous stomatal structure, and solid ventral tissue with striate wall markings. f. Stoma and surrounding cells in surface view. g. Ventral scale.

Fig. 27. *Reboulia hemisphaerica.* Female receptacles, typically with 5–7 shallow lobes. The thallus surface is smooth-textured and, at least when dry, appears to lack areolation and stomatal structures. (Photo by E. B. Mains.)

150

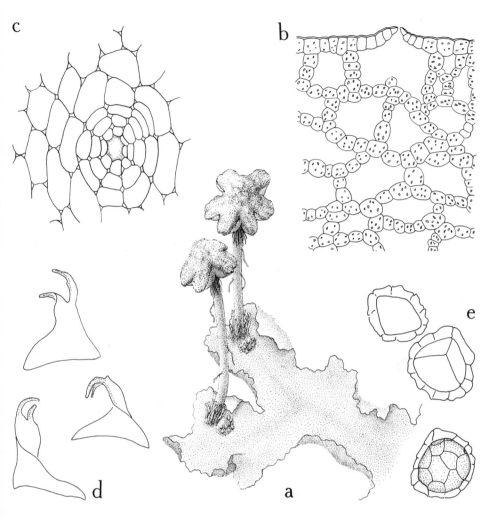

Fig. 28. *Reboulia hemisphaerica.* a. Thallus, with sessile male and stalked female receptacles. b. Portion of the thallus in section. c. Upper epidermal cells, with pore; note thickened cell corners. d. Ventral scales. e. Spores.

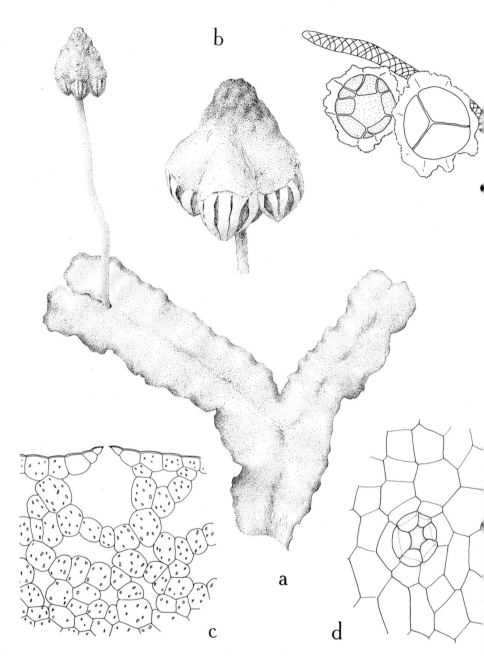

Fig. 29. *Asterella tenella*. a. Thallus, with female receptacle. b. Female receptacle, enlarged, with deeply cleft pseudoperianths. c. Portion of the thallus in section. d. Upper epidermal cells, with pore. e. Spores and elater.

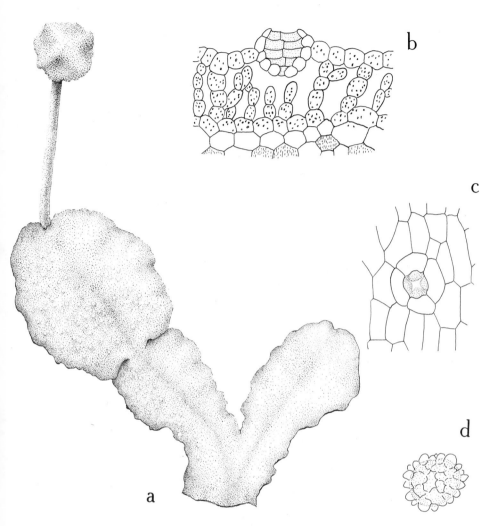

Fig. 30. *Preissia quadrata.* a. Thallus, with female receptacle. b. Portion of thallus in section, showing a barrel-shaped stomatal structure. c. Upper epidermal cells, with pore. d. Spore.

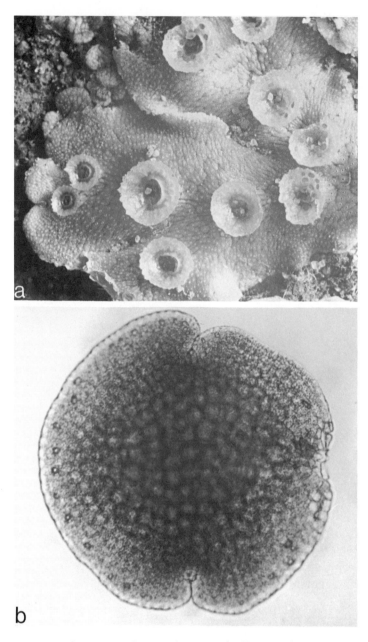

Fig. 31. *Marchantia polymorpha.* a. Thallus with gemmae cups (photo by E. B. Mains). b. Gemmae with growing points in the notch at either end; the smaller cells around the periphery are oil cells, each containing a single large oil body (photo by Jane Taylor).

Fig. 32. *Marchantia polymorpha.* a. Female plants (most commonly with 9-rayed receptacles). In the lower right-hand corner is a discoid male receptacle. b. Female plants with dehisced capsules exposing cottony yellow masses of elaters and spores at the undersurfaces of receptacles. Rhizoids are seen on stalks of receptacles and also on undersurfaces of thalli. (Photos by Larry Mellichamp.)

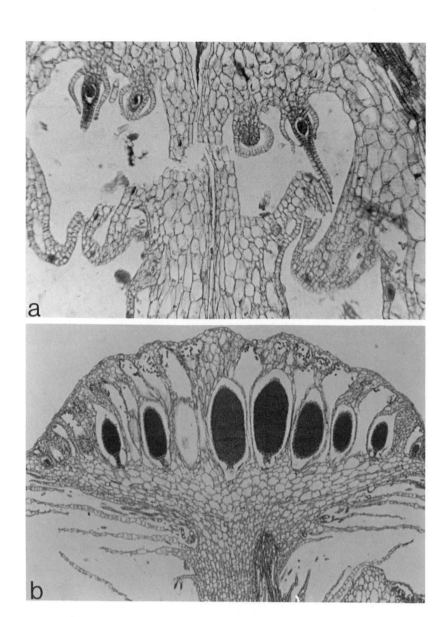

Fig. 33a–b. *Marchantia polymorpha.* a. Section through a female receptacle, showing a pseudoperianth around each archegonium. Groups of archegonia and pseudoperianths are enclosed by an involucre (photo by Gordon McBride). b. Section through a discoid male receptacle, showing antheridia singly produced in air chambers.

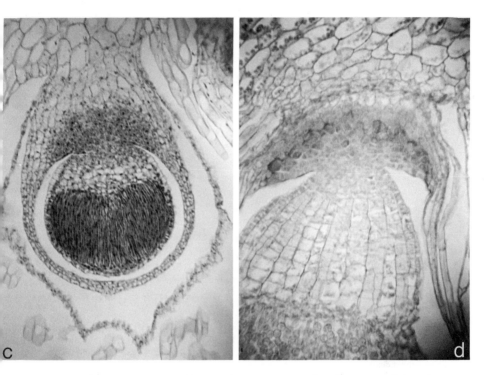

Fig. 33c–d. *Marchantia polymorpha.* c. Longitudinal section through a sporophyte, showing foot, seta, and capsule. The sporophyte is enclosed in a calyptra and, external to that, a pseudoperianth. (The involucre, a fringed structure surrounding a group of archegonia or sporophytes, is not shown.) Spores and elaters are vertically aligned within the capsule. d. Longitudinal section through foot and seta.

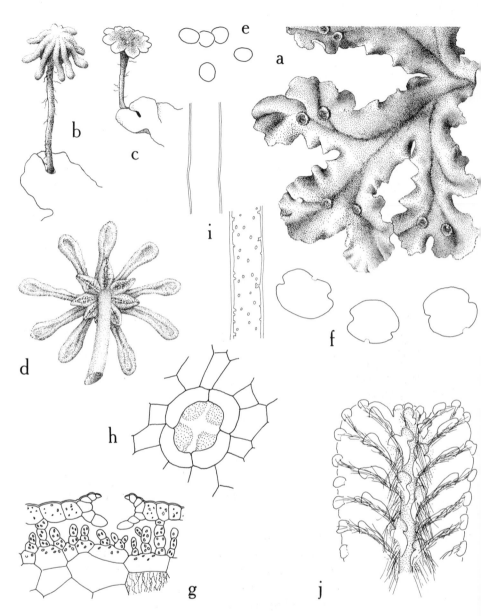

Fig. 34. *Marchantia polymorpha.* a. Thallus with gemmae cups. b. Female receptacle. c. Male receptacle. d. Female receptacle, undersurface, showing 9 rays alternating with 8 involucres. e. Spores. f. Gemmae. g. Portion of thallus in section, showing air chamber and pore structure. h. Upper epidermis, with pore. i. Portion of smooth and pegged rhizoids. j. Ventral surface of thallus lobe, showing scales in three rows on either side of the midrib and strands of rhizoids.

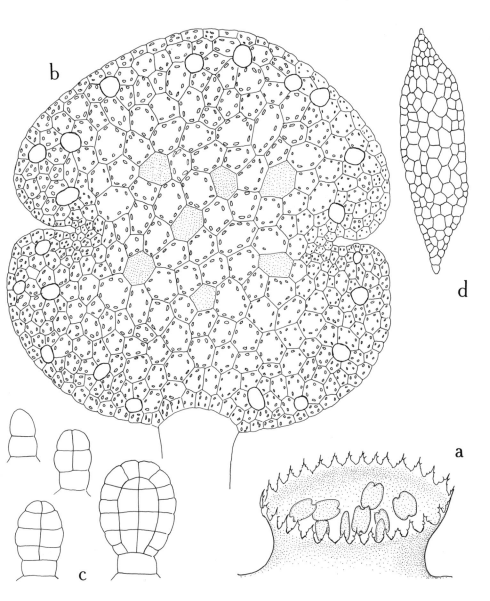

Fig. 35. *Marchantia polymorpha*. a. Gemmae-cup, showing laciniate margin. b. Gemma, showing rounded oil cells and rhizoid initials scattered in the interior (redrawn from Kny, *Botanische Wandtafeln*, 1890). c. Stages in the early development of a gemma. d. Longitudinal section through a gemma. (c, d, redrawn from Müller, *Lebermoose Europas*).

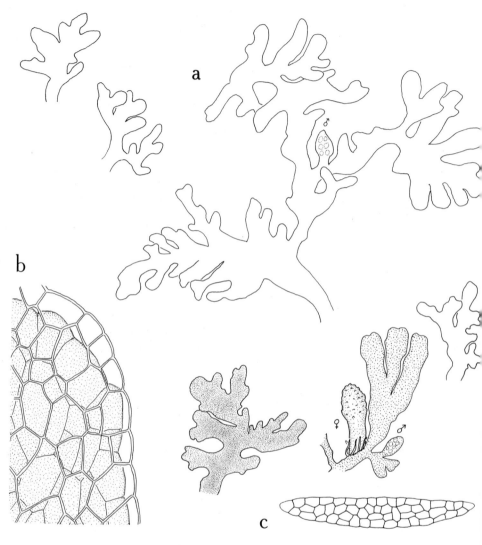

Fig. 36. *Riccardia latifrons.* a. A variety of thallus shapes, two of them with reproductive structures. b. Margin of thallus, showing at most a single row of unistratose cells. c. Thallus in section.

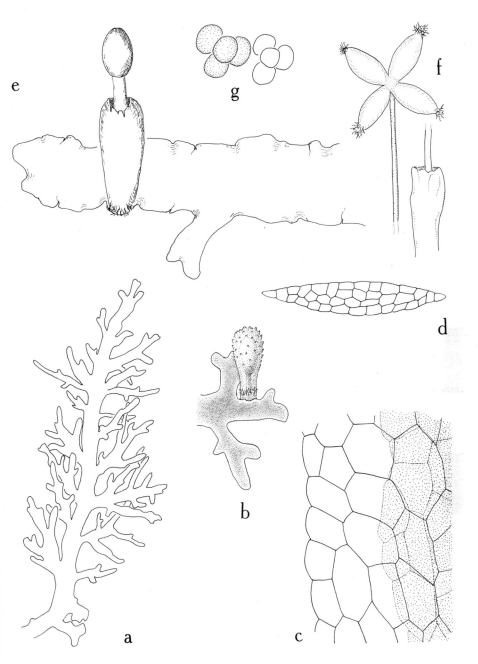

Fig. 37. *Riccardia multifida.* a. Thallus. b. Portion of thallus, with involucral fringe and calyptra. c. Thallus margin, showing about 2–3 rows of unistratose cells. d. Thallus in section. *Aneura pinguis.* e. Thallus, with calyptra and sporophyte. f. Dehiscence of capsule, showing four valves, each with an apical tuft of elaters. g. Deeply 4-lobed spore mother cells.

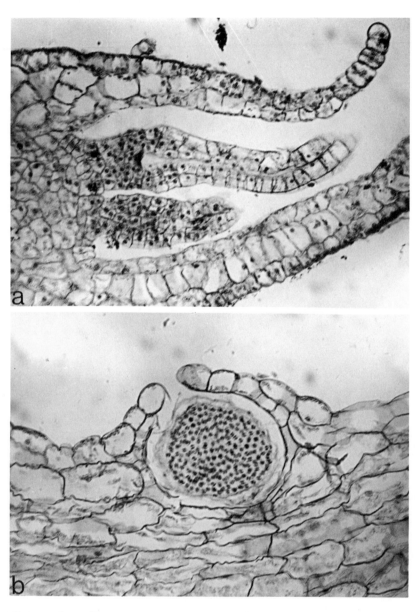

Fig. 38a–b. *Pellia epiphylla.* a. Archegonia produced behind the growing point and protected by an involucral flap. b. Section through thallus, showing an antheridial chamber.

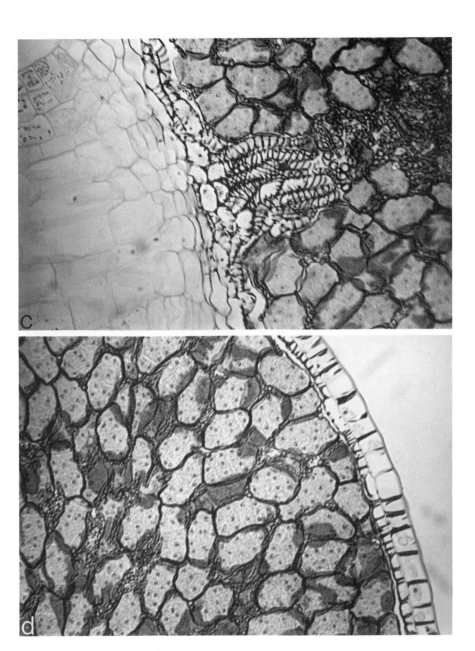

Fig. 38c–d. *Pellia epiphylla.* c. Base of capsule showing a tuft of elaters. d. Portion of capsule in section, showing a wall of 2 layers of cells. Spore mother cells and elaters are beginning to differentiate.

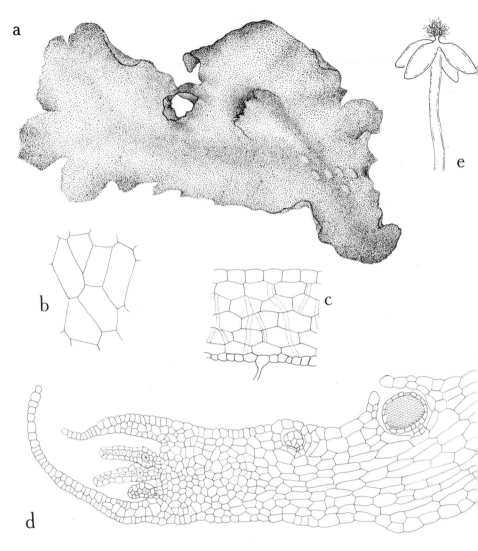

Fig. 39. *Pellia epiphylla.* a. Thallus, with a flaplike involucre and, behind it, a group of antheridial chambers. (The tear behind the growing point is only incidental.) b. Upper epidermis. c. Thallus middle in longitudinal section, showing thickened bands. d. Longitudinal section through thallus, showing archegonia under an involucral flap and 2 antheridial chambers. e. Capsule in dehiscence, showing a tuft of elaters at the base.

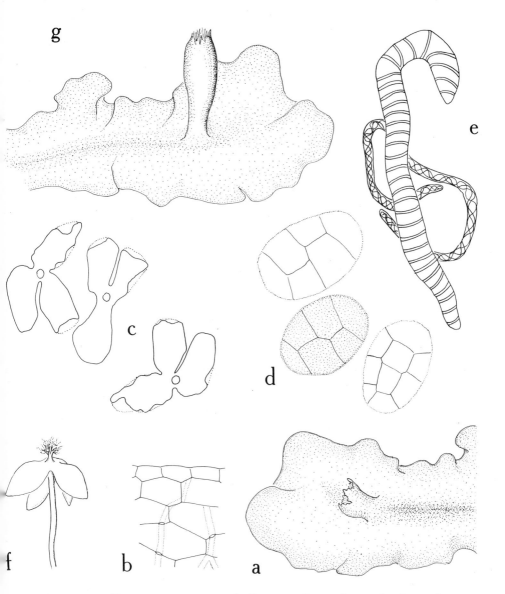

Fig. 40. *Pellia neesiana.* a. Thallus, with a short, horizontal perigynium irregularly lobed and crenate at the mouth. b. Thallus middle in longitudinal section, showing thickened bands. c. Four-lobed spore mother cells as seen in section (only 3 lobes showing). d. Spores. e. Free elater and a much stouter fixed elater. f. Capsule in dehiscence, showing a basal tuft of elaters attached to elaterophores. *Pellia megaspora.* g. Thallus, showing an erect, cylindric perigynium with a ciliate mouth.

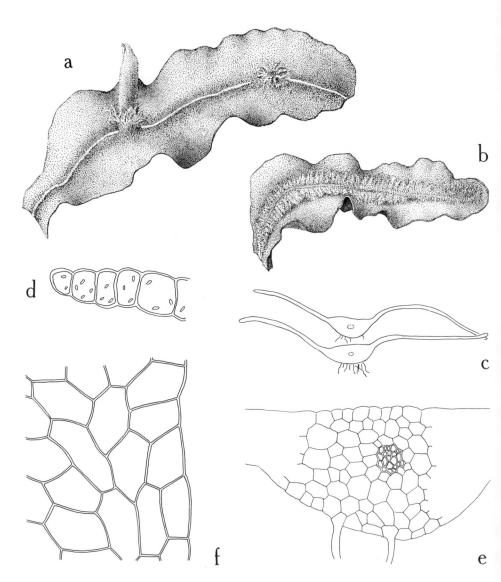

Fig. 41. *Pallavicinia lyellii.* a. Female plant. b. Male plant (smaller than the female). c. Thallus sections. d. Unistratose thallus margin in section. e. Thallus middle in section, showing a central strand. f. Epidermal cells of thallus.

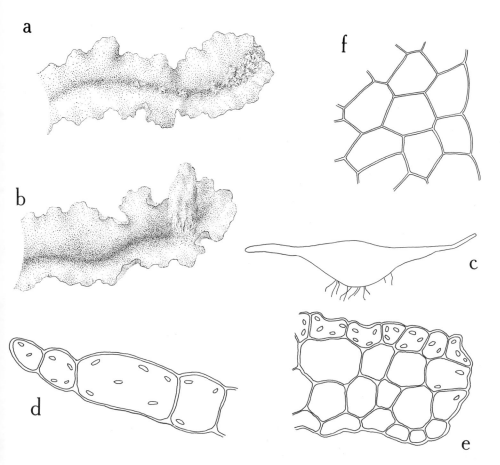

Fig. 42. *Moerckia hibernica.* a. Male plant. b. Female plant. c. Thallus in section, with thickened wings. d, e. Margins of thallus, both unistratose and multistratose. f. Epidermal cells of thallus.

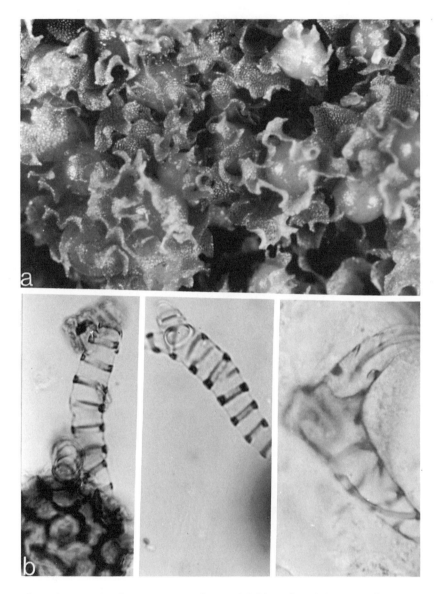

Fig. 43. *Fossombronia cristula.* a. Habit, showing capsules surrounded by ruffled perigynia. b. Elaters, stout and irregular in shape, with annular or unispiral wall thickenings.

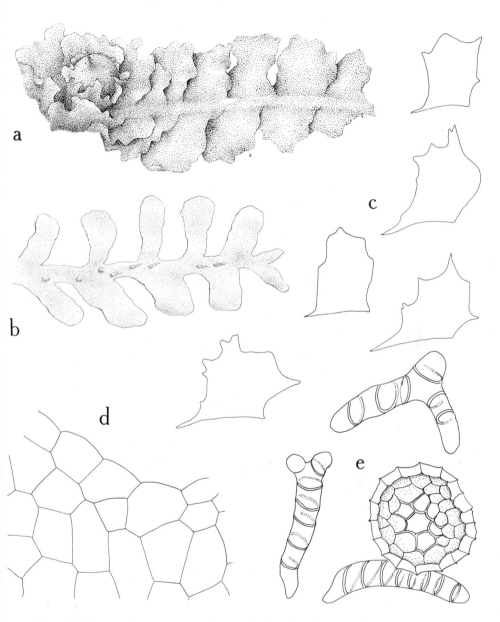

Fig. 44. *Fossombronia cristula.* a. Thallus, with capsule surrounded by a bowl-shaped involucre with a ruffled margin. b. Thallus, with less crowded leaves, bearing dorsal archegonia and, behind them, antheridia. c. Leaves of etiolated plants grown in culture, showing irregular shapes and marginal teeth and lobes. d. Cells at margin of leaf. e. Spore and elaters.

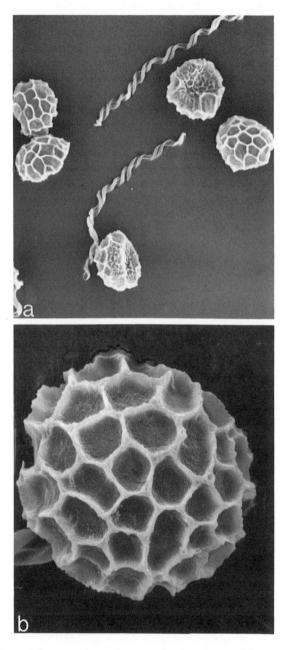

Fig. 45. *Fossombronia foveolata.* a, b. Spores and bispirally thickened elaters.

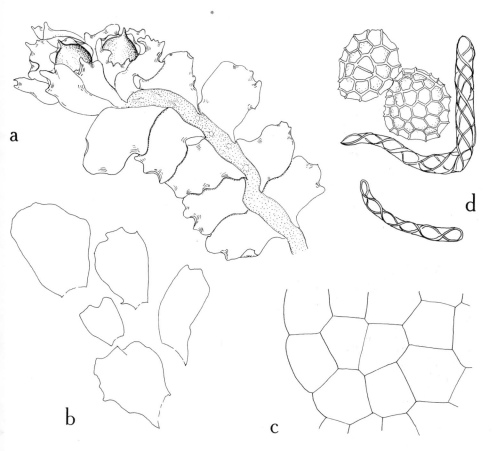

Fig. 46. *Fossombronia foveolata.* a. Thallus, with dorsal sex organs and 2 capsules enclosed in ruffled, bowl-shaped perigynia. b. Leaves. c. Cells at leaf margin. d. Spores and elaters.

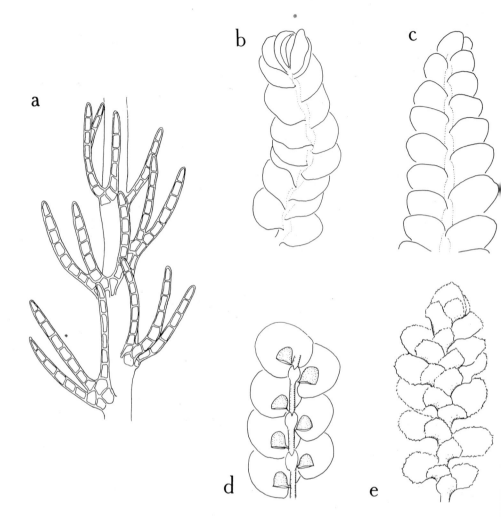

Fig. 47. Basic arrangements and forms of leaves. a. Transversely inserted leaves of *Blepharostoma*. b. Succubous leaves of *Mylia*. c. Incubous leaves of *Calypogeja*. d. Complicate-bilobed leaves of *Frullania*, ventral view. e. Complicate-bilobed leaves of *Scapania*, dorsal view.

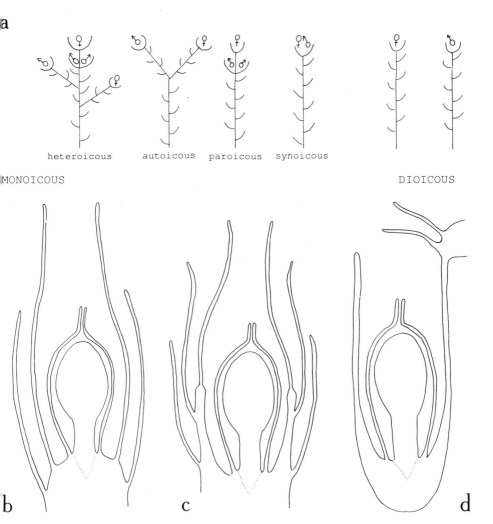

heteroicous   autoicous paroicous synoicous

MONOICOUS   DIOICOUS

b   c   d

Fig. 48. Inflorescence types. a. Dioicous and monoicous expressions of sexuality. b–d. Sections through female inflorescences in diagram. b. *Lophocolea*, showing a sporophyte inside a calyptra which is surrounded by a perianth and that in turn by involucral bracts. c. *Marsupella*, showing a sporophyte inside a calyptra that is surrounded by a perigynium (of hollowed-out stem tissue) subtending a perianth and bearing bracts in 2 whorls. d. *Calypogeja*, showing the sporophyte inside a calyptra that is sunken into a pendent perigynium (of hollowed-out stem tissue), often called a marsupium; two bracts are shown on the horizontal stem above the marsupium. [Adapted from Evans, *The Bryologist*, 1905.]

173

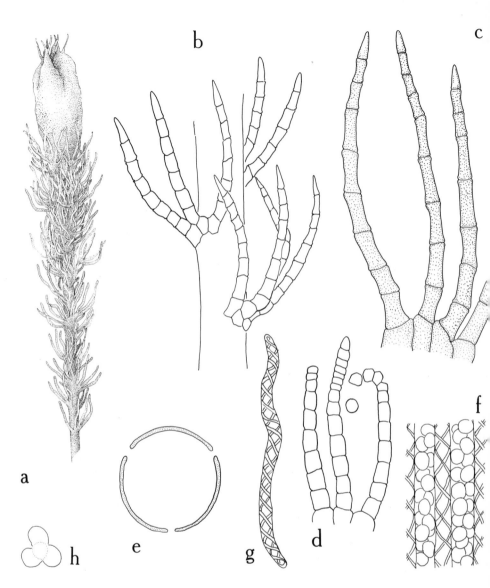

Fig. 49. *Blepharostoma trichophyllum*. a. Habit. b. Portion of stem showing transverse leaf insertion. c. Four-lobed leaf. d. Gemmae-bearing upper leaf. e. Diagram showing how the perianth was formed by the evolutionary coalescence of 2 leaves and an underleaf. f. Spores and elaters formed in alternating vertical rows within the capsule. g. Elater. h. Deep lobing of a spore mother cell prior to meiosis. (f–h, after Hollensen, *Journal of the Hattori Botanical Laboratory*, 1964.)

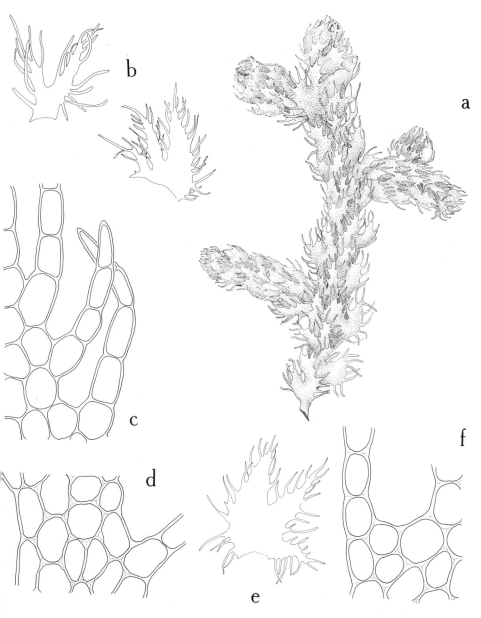

Fig. 50. *Ptilidium pulcherrimum.* a. Habit. b. Leaves. c, d. Leaf cells showing variations in trigone development. *P. ciliare.* e. Leaf. f. Leaf cells.

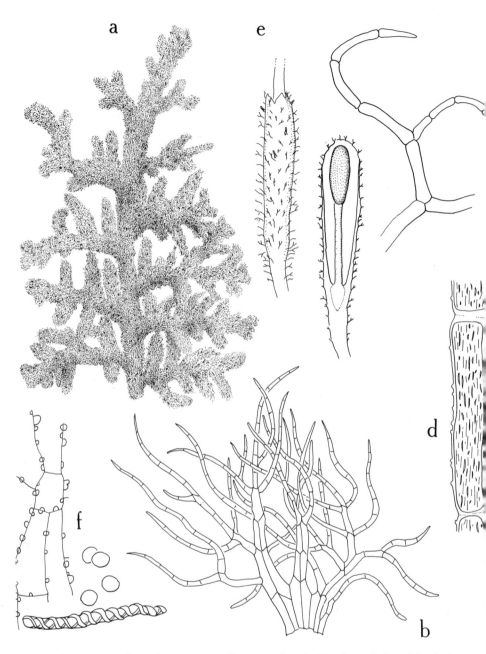

Fig. 51. *Trichocolea tomentella*. a. Habit. b. Leaf. c. Cells of leaf. d. Cell of leaf, enlarged, showing punctate striae. e. Marsupium, aspect and sectional views. f. Epidermal cell of the capsule wall, with spores and bispiral elater.

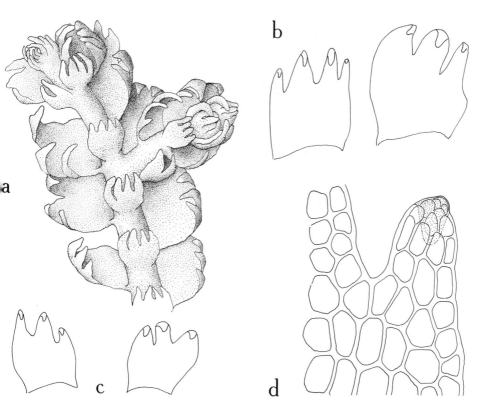

Fig. 52. *Lepidozia reptans.* a. Ventral view of portion of plant. b. Leaves. c. Underleaves. d. Leaf cells.

Fig. 53. *Bazzania trilobata.* a, b. Habits, showing dichotomous forking and ventral branchlets (photos by David Bay).

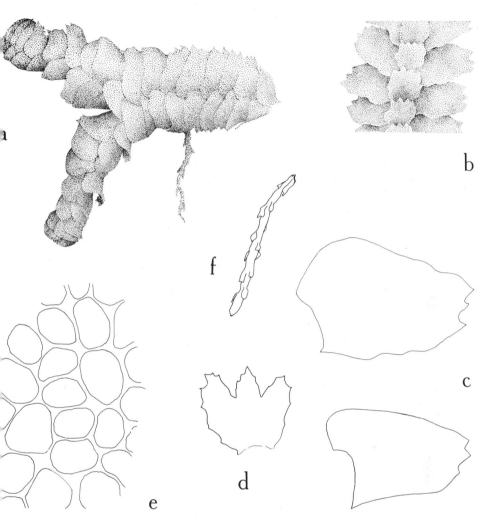

Fig. 54. *Bazzania trilobata.* a. Habit. b. Ventral view of stem, showing underleaves. c. Leaves. d. Underleaf. e. Median leaf cells. f. Ventral branchlet, with reduced leaves.

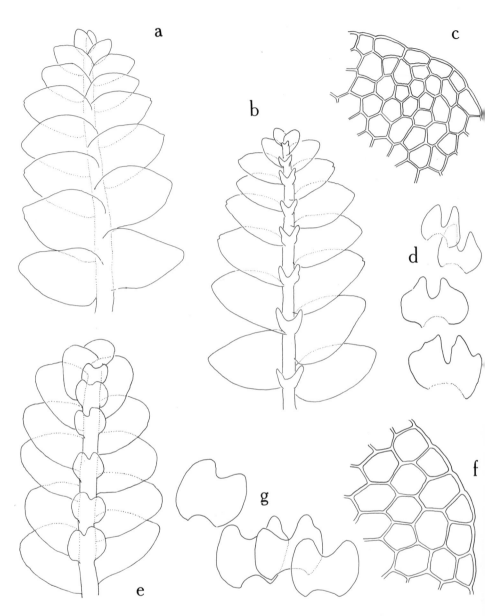

Fig. 55. *Calypogeja fissa.* a, b. Habits, dorsal and ventral views. c. Cells at leaf apex. d. Underleaves. *Calypogeja muelleriana.* e. Habit, ventral view. f. Cells at leaf apex. g. Underleaves.

Fig. 56. *Calypogeja muelleriana.* Habit, showing the incubous, unlobed leaves that characterize the genus.

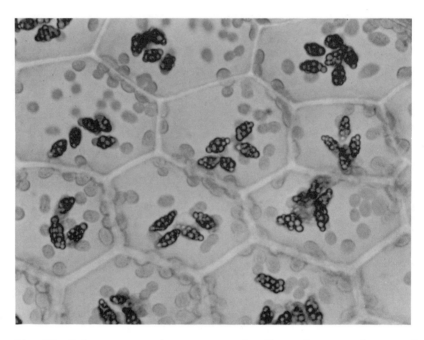

Fig. 57. *Calypogeja trichomanis.* Leaf cells with grape-cluster oil bodies (photo by L. E. Anderson).

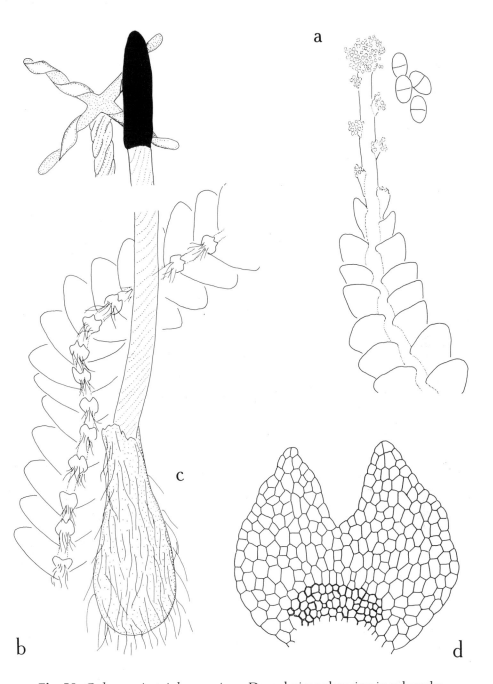

Fig. 58. *Calypogeja trichomanis*. a. Dorsal view showing incubously overlapping leaves and apical clusters of gemmae. b. Ventral view showing rhizoids attached to lower part of underleaves. c. Marsupium, young sporophyte, and dehisced capsule with twisted valves. (Adapted from Casares-Gil, *Flora Ibérica*, 1919.) d. Underleaf, with a basal zone of rhizoid initials. (Redrawn from Schuster, *New Manual of Bryology*, 1984.)

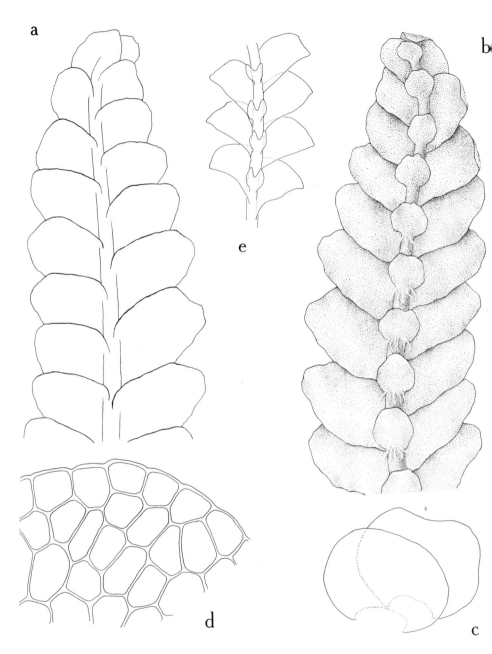

Fig. 59. *Calypogeja integristipula.* a, b. Habits, dorsal and ventral views. c. Underleaves. d. Cells at apex of leaf. *Calypogeja sphagnicola.* e. Portion of plant, ventral view.

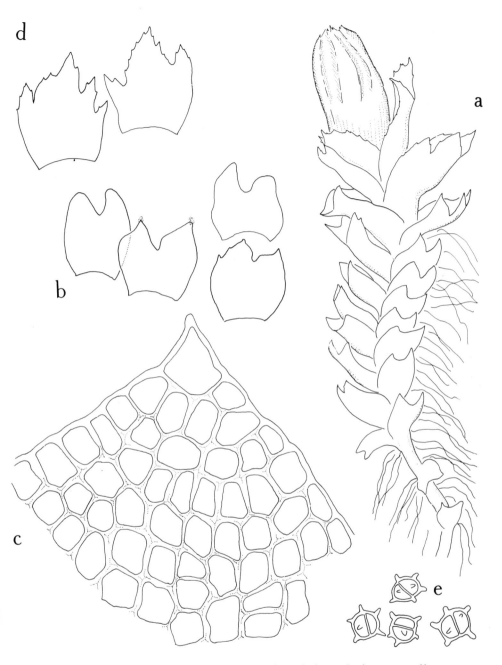

Fig. 60. *Lophozia bicrenata*. a. Habit (adapted from Müller, *Lebermoose Europas*). b. Leaves. c. Leaf apex. d. Bracts of the involucre. e. Gemmae.

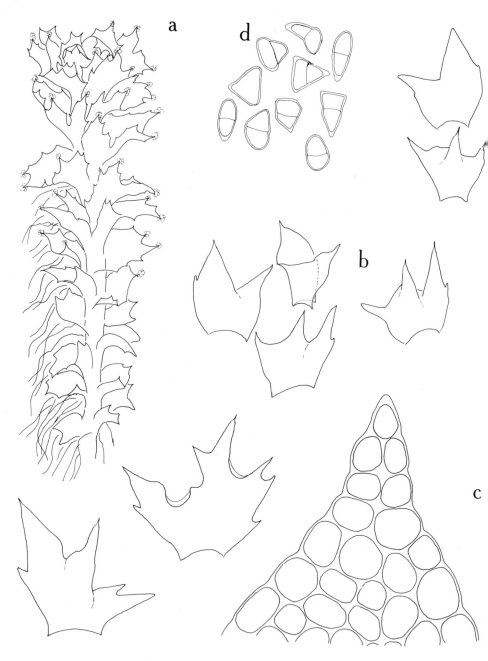

Fig. 61. *Lophozia incisa.* a. Habit (adapted from Müller, *Lebermoose Europas*). b. Leaves and bracts. c. Cells of leaf apex. d. Gemmae.

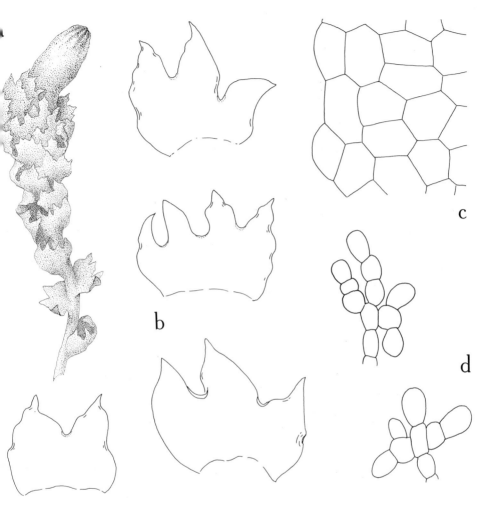

Fig. 62. *Lophozia capitata.* a. Habit. b. Leaves. c. Cells at leaf margin.
d. Gemmae.

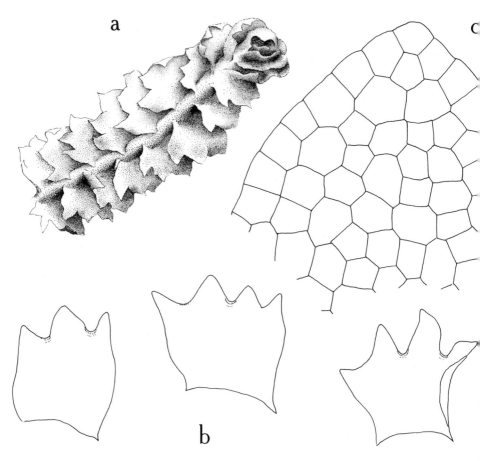

a

c

b

Fig. 63. *Lophozia barbata*. a. Habit. b. Leaves. c. Leaf cells.

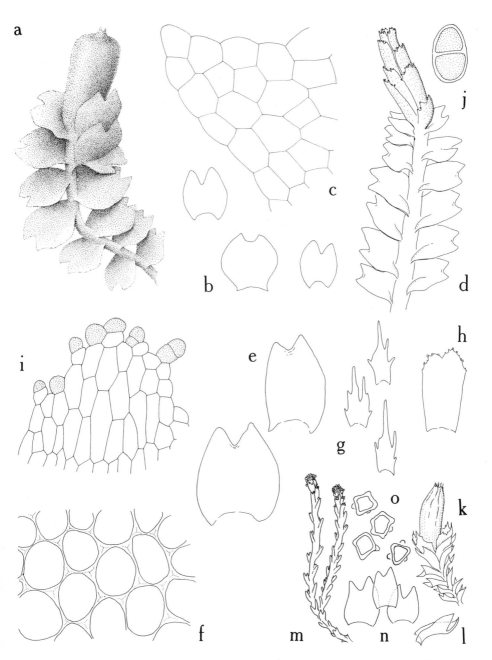

Fig. 64. *Lophozia badensis.* a. Habit. b. Leaves. c. Leaf cells. *Lophozia heterocolpos.* d. Habit. e. Leaves. f. Leaf cells. g. Underleaves. h. Gemmae-bearing leaf. i. Apex of gemmae-bearing leaf. j. Gemma. *Anastrophyllum hellerianum.* k. Habit of plant, with perianth. l. Leaf of fertile plant. m. Habit of gemmiparous shoots. n. Leaves of gemmiparous plant. o. Gemmae.

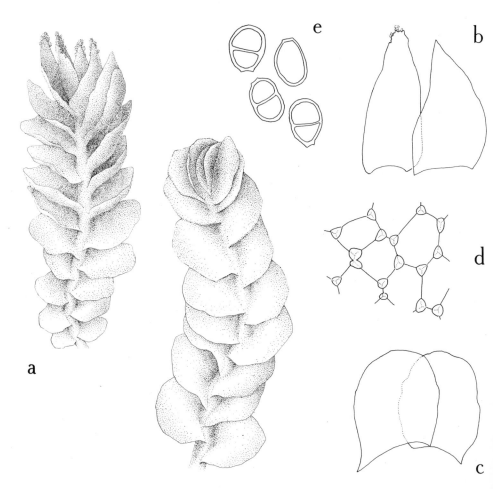

Fig. 65. *Mylia anomala.* a. Habits, with and without gemmae. b. Upper leaves, often gemmae-bearing at the tips. c. Lower leaves. d. Leaf cells. e. Gemmae.

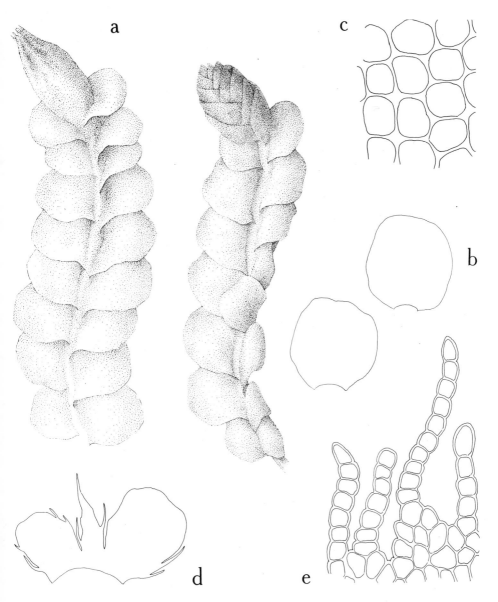

Fig. 66. *Jamesoniella autumnalis.* a. Habits, showing male and female inflorescences. b. Leaves. c. Leaf cells. d. Involucre opened out to show bracts and bracteole. e. Cilia at mouth of perianth.

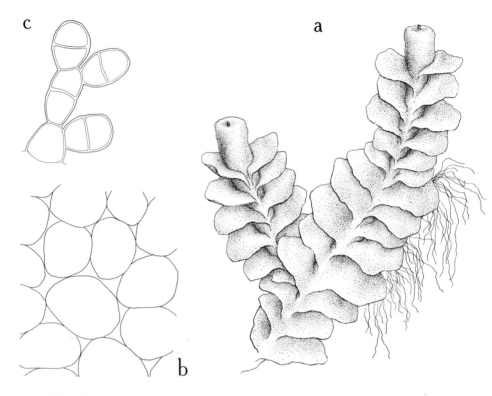

Fig. 67. *Jungermannia leiantha*. a. Habit. b. Leaf cells. c. Gemmae.

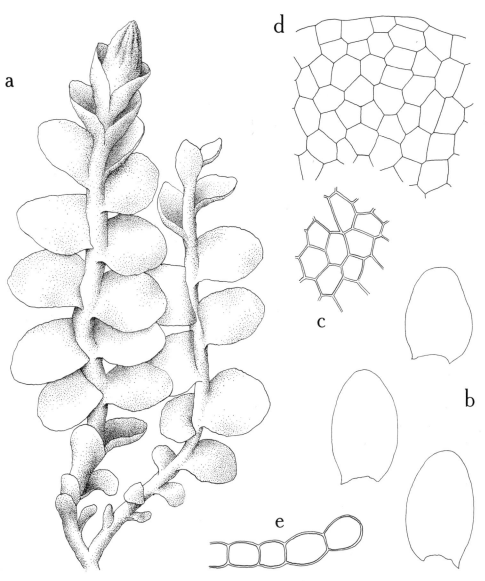

Fig. 68. *Solenostoma pumilum.* a. Habit. b. Leaves. c. Median leaf cells. d. Cells at leaf apex. (The thickness of cell walls is subject to variation.) e. Leaf margin in section.

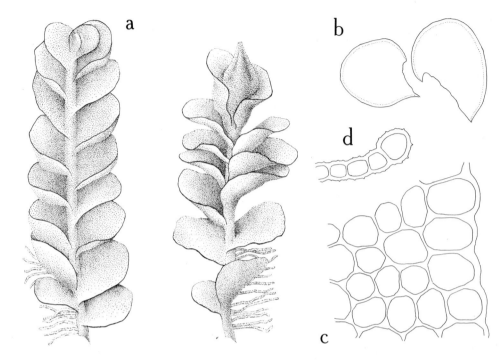

Fig. 69. *Solenostoma crenuliforme*. a. Habits. b. Leaves. c. Cells at leaf apex. d. Leaf margin in section.

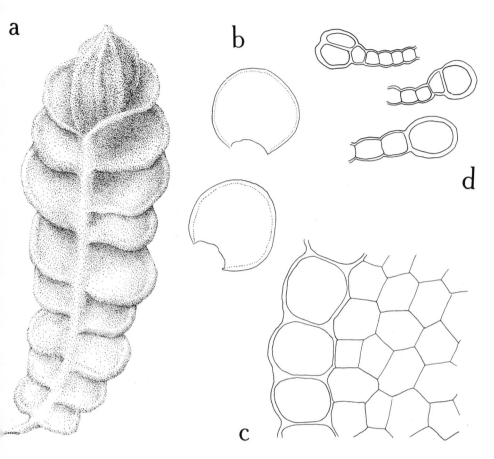

Fig. 70. *Solenostoma gracillimum*. a. Habit. b. Leaves. c. Cells at leaf margin. d. Leaf margins in section.

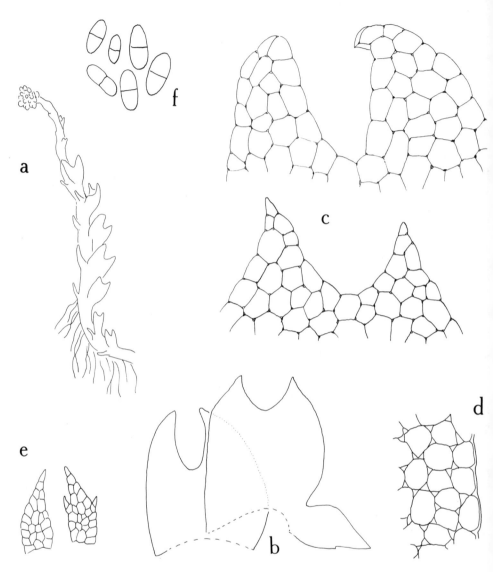

Fig. 71. *Harpanthus drummondii*, small gemmiparous form. a. Habit (adapted from Steere, *Liverworts of Southern Michigan*). b. Leaves, one with attached underleaf. c. Leaf tips. d. Leaf cells. e. Underleaves. f. Gemmae.

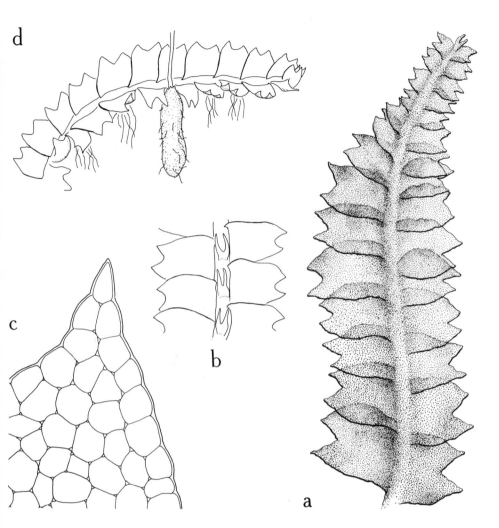

Fig. 72. *Geocalyx graveolens.* a. Habit. b. Ventral view of plant, showing underleaves. c. Cells at leaf apex. d. Leafy shoot with ventral marsupium (redrawn from Müller, *Lebermoose Europas*).

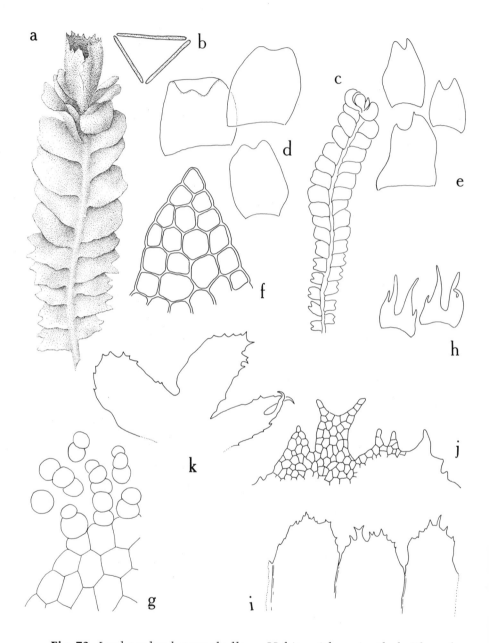

Fig. 73. *Lophocolea heterophylla.* a. Habit, with perianth. b. Plan of
perianth formed by the fusion of 2 leaves and 1 underleaf, with keels
at places of fusion. c. Plant with bidentate lower leaves and truncate
to emarginate upper leaves. d. Slightly indented leaves of upper portion of a fertile stem. e. Bidentate juvenile leaves. f. Leaf cells. g. Gemmae at leaf margins (very rare); the slight thickenings of cell walls not
indicated. h. Underleaves. i. Mouth of perianth. j. Toothing at mouth
of perianth. k. Involucre, spread out to show 2 bracts and 1 bracteole.

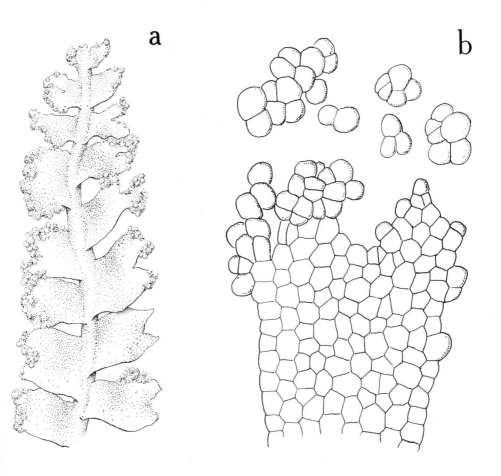

Fig. 74. *Lophocolea minor.* a. Habit. b. Leaf, with gemmae (adapted from Müller, *Lebermoose Europas*).

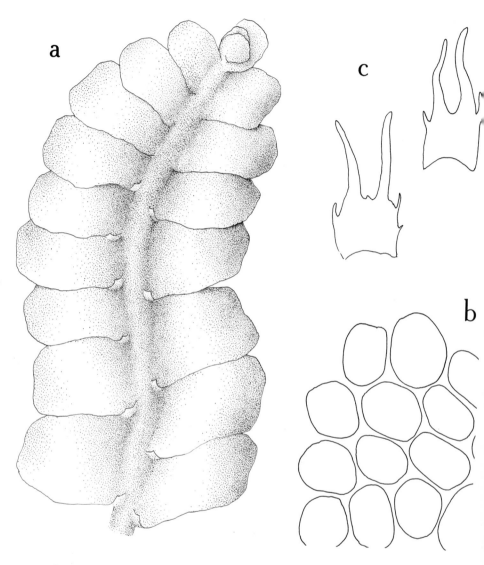

Fig. 75. *Chiloscyphus pallescens.* a. Habit, showing leaflike antheridial bracts (with basal lobules). b. Median leaf cells. c. Underleaves.

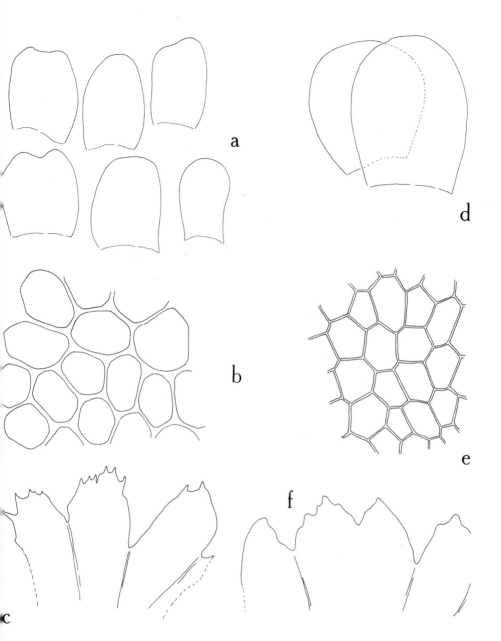

Fig. 76. *Chiloscyphus pallescens* in comparison with *C. polyanthos.*
a–c. *C. pallescens*, leaves, leaf cells, and perianth opened out. d–f. *C. polyanthos*, leaves, leaf cells, and perianth opened out.

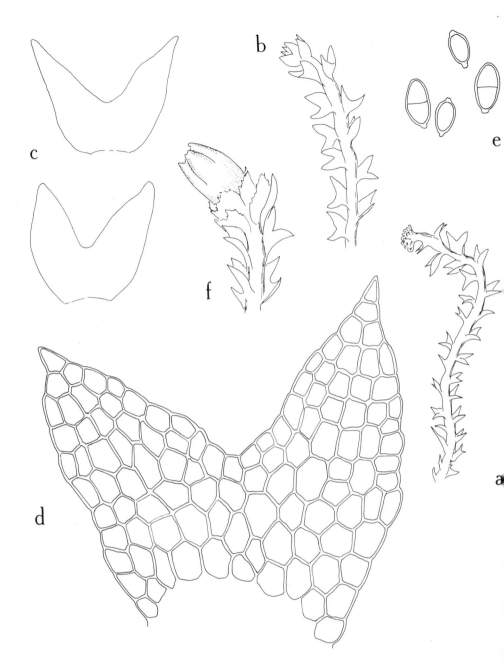

Fig. 77. *Cephaloziella hampeana.* a, b. Habits, with and without gemma clusters. c. Leaves. d. Leaf cells. e. Gemmae. f. Portion of plant, with perianth.

Fig. 78. *Cephaloziella rubella*. a. Gemmiparous plant. b. Portion of female plant. c. Portion of male plant. d. Gemmae. e. Leaf outlines. f. Leaves with cells. g. Involucre opened out to show 2 bracts and 1 bracteole.

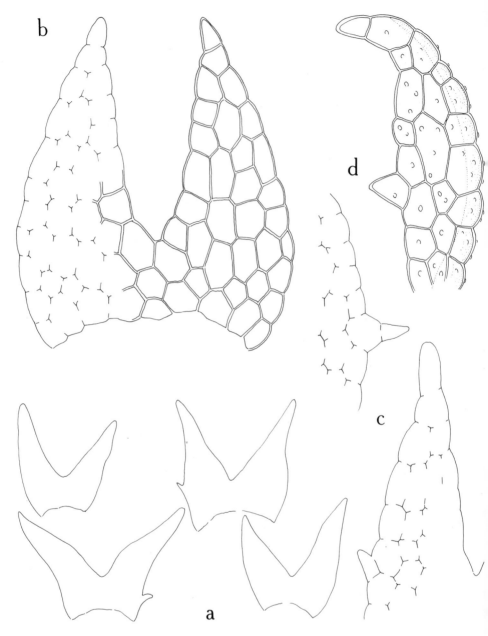

Fig. 79. *Cephaloziella spinigera.* a. Leaves. b. Leaf with areolation partially shown. c. Leaf margins, both entire and toothed. d. Leaf lobe partly in profile showing papillae at back.

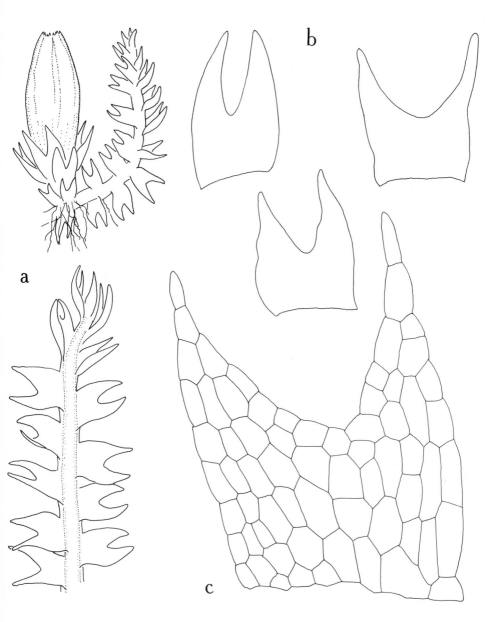

Fig. 80. *Cephalozia bicuspidata.* a. Habits, with and without perianth (at different magnifications). b. Leaves. c. Leaf, with cells.

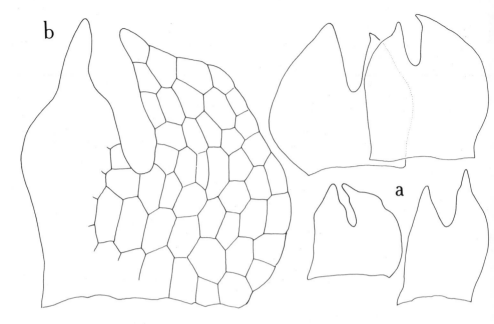

Fig. 81. *Cephalozia pleniceps.* a. Leaves. b. Leaf with cells.

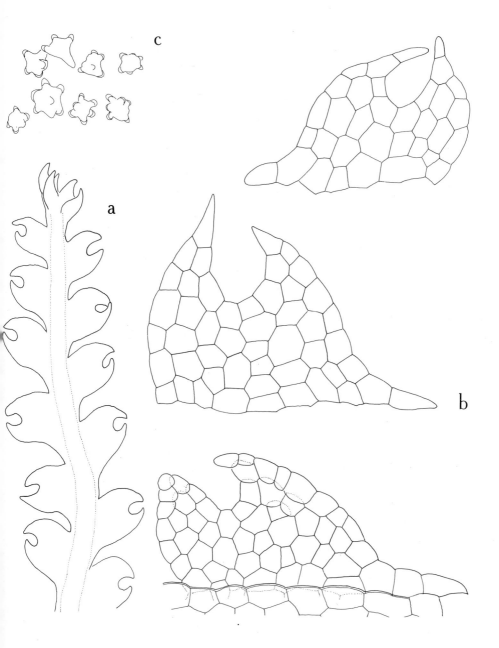

Fig. 82. *Cephalozia lunulifolia.* a. Habit. b. Leaves with cells. c. Gemmae.

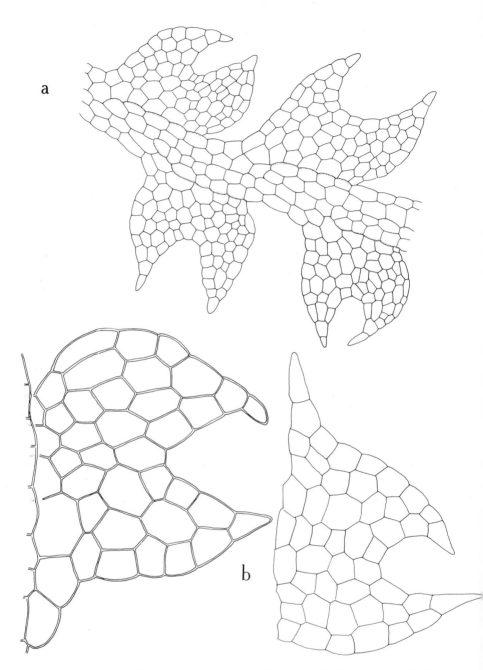

a

b

Fig. 83. *Cephalozia connivens.* a. Portion of plant (adapted from Buch, *Annales Bryologici*, 1930). b. Comparison of leaves and leaf cells of *C. connivens* (left) with those of *C. lunulifolia* (right).

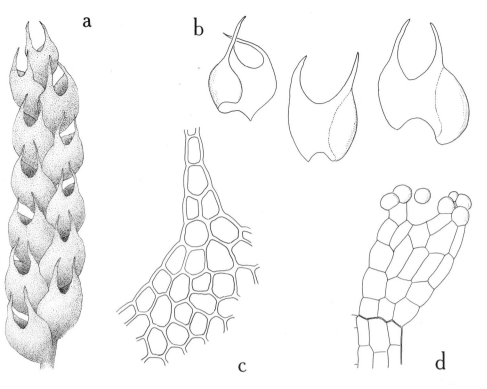

Fig. 84. *Nowellia curvifolia.* a. Habit. b. Leaves. c. Leaf cells. d. Reduced, gemma-bearing leaf.

c

a

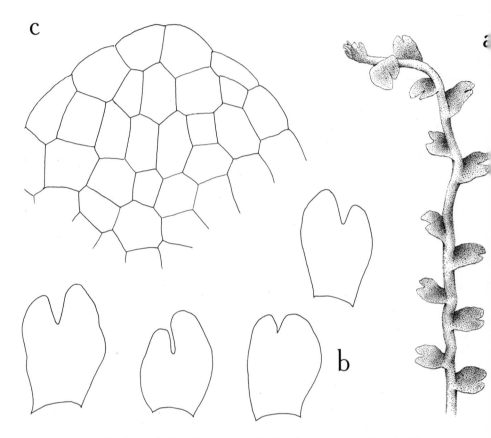

Fig. 85. *Cladopodiella fluitans*. a. Habit. b. Leaves. c. Leaf cells.

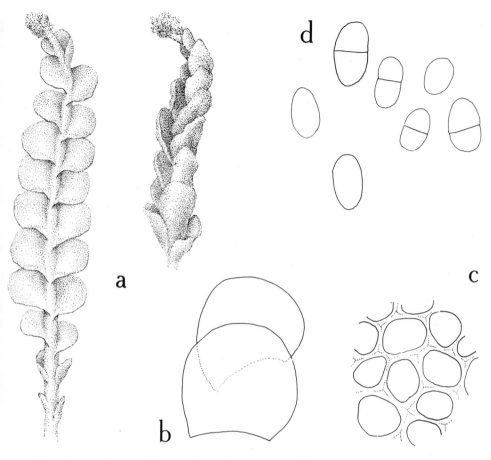

Fig. 86. *Odontoschisma denudatum.* a. Habits, moist and dry. b. Leaves. c. Leaf cells. d. Gemmae.

Fig. 87. *Plagiochila porelloides.* Habit of living plants. The concavity of the lower portion of the leaves continued down the stem as a decurrency gives a distinctive aspect.

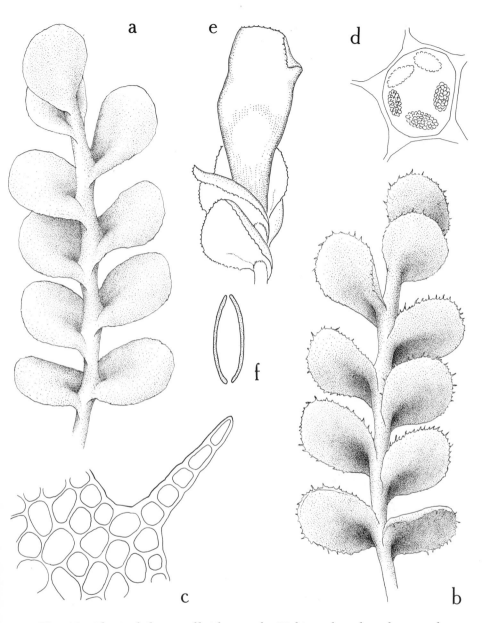

Fig. 88. *Plagiochila porelloides.* a, b. Habits, dorsal and ventral views, showing variation in marginal toothing. c. Leaf cells. d. Leaf cell, with oil bodies. e. Perianth as seen from the side. f. Plan of the laterally compressed perianth formed by the fusion of 2 lateral leaves, no underleaf.

213

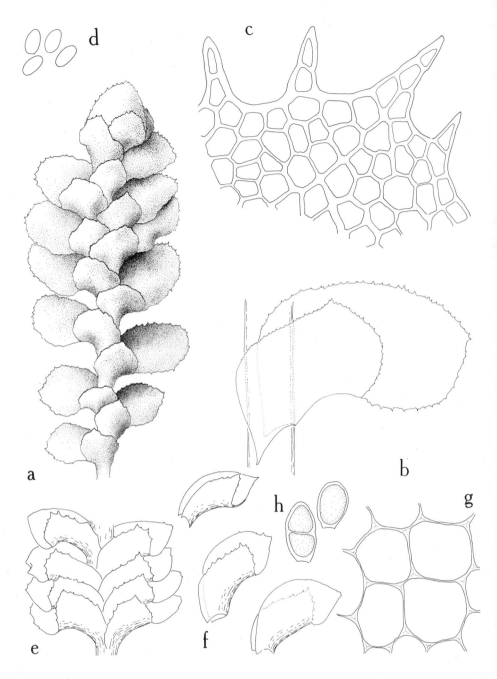

Fig. 89. *Scapania nemorea.* a. Habit, dorsal view. b. Leaf. c. Cells at leaf margin. d. Gemmae. *Scapania saxicola.* e. Portion of leafy stem. f. Leaves. g. Leaf cells. h. Gemmae.

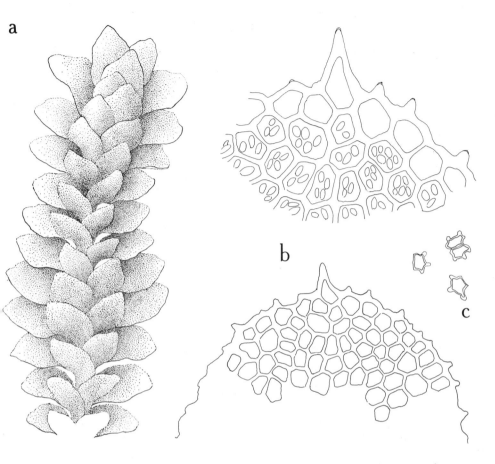

Fig. 90. *Diplophyllum apiculatum.* a. Habit. b. Leaf tips, drawn with and without papillae and at different magnifications. c. Gemmae.

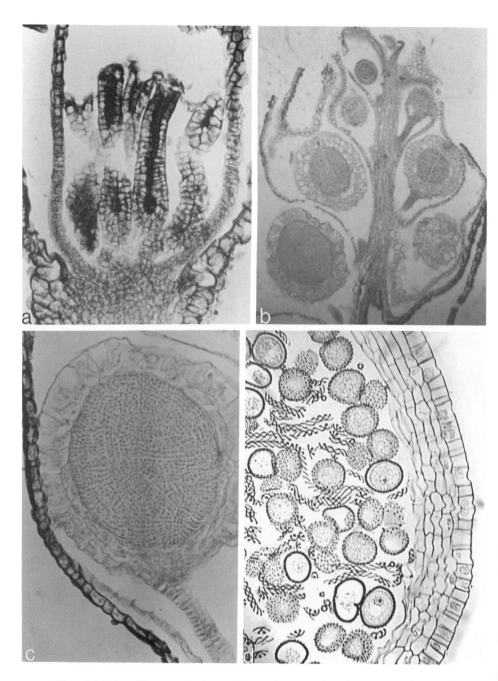

Fig. 91. *Porella.* a. Archegonia in longitudinal section. b. Male inflorescence. The antheridia are produced singly in the axils of modified leaves. The 2-layered jacket is distinctive. The globose shape is characteristic of leafy liverworts. c. A single antheridium. d. Portion of the capsule in section, showing spores and elaters bounded by a multistratose wall.

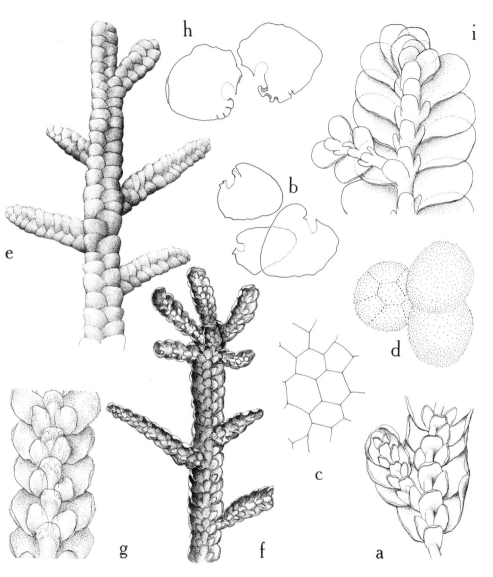

Fig. 92. *Porella platyphylla.* a. Portion of plant, ventral view. b. Dorsal lobe of 3 leaves. c. Leaf cells. d. Spores. *P. platyphylloidea.* e, f. Habits, dorsal and ventral views. g. Portion of plant, ventral view, for comparison with corresponding figure (a) of *P. platyphylla.* h. Dorsal lobe of 2 leaves. *P. pinnata.* i. Portion of plant, ventral view.

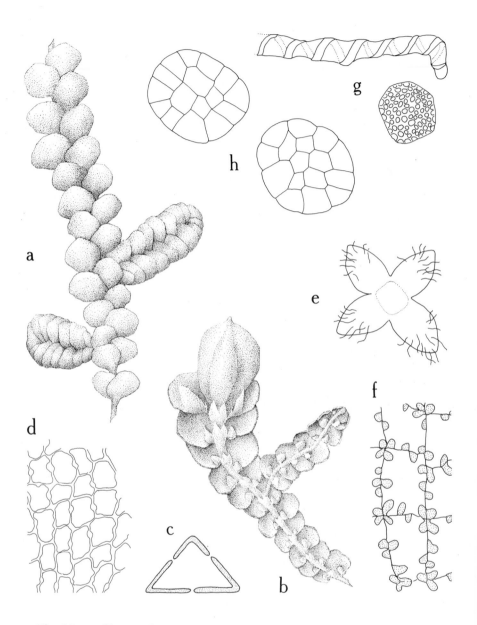

Fig. 93. *Frullania eboracensis.* a. Male plant, dorsal view. b. Female plant, ventral view. c. Plan of the perianth formed by the fusion of 2 lateral leaves and 1 underleaf. d. Leaf cells, with intermediate thickenings. e. Capsule after dehiscence. f. Epidermal cells of capsule wall. g. Portion of elater with spore. h. Large, multicellular (precociously germinated) spores that have shed their spore wall.

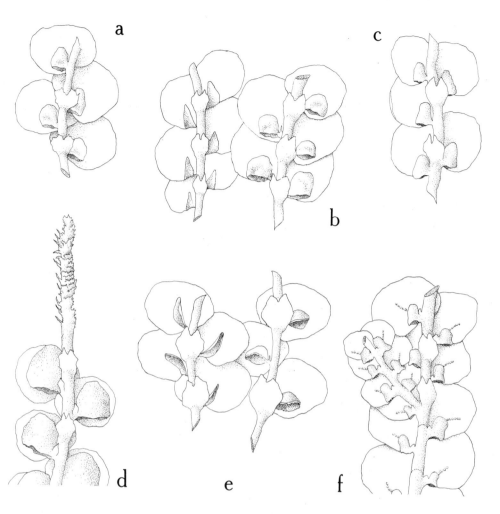

Fig. 94. Species of *Frullania* in habit comparisons. a. *F. brittoniae*. b. *F. inflata*. c. *F. eboracensis*. d. *F. bolanderi*. e. *F. riparia*. f. *F. asagrayana*.

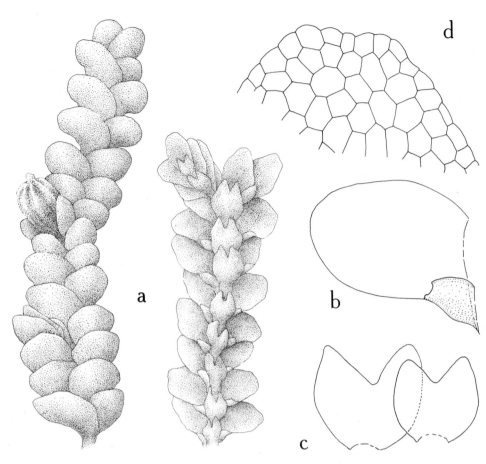

Fig. 95. *Lejeunea cavifolia.* a. Habits, dorsal and ventral views. b. Leaf. c. Underleaves. d. Leaf cells.

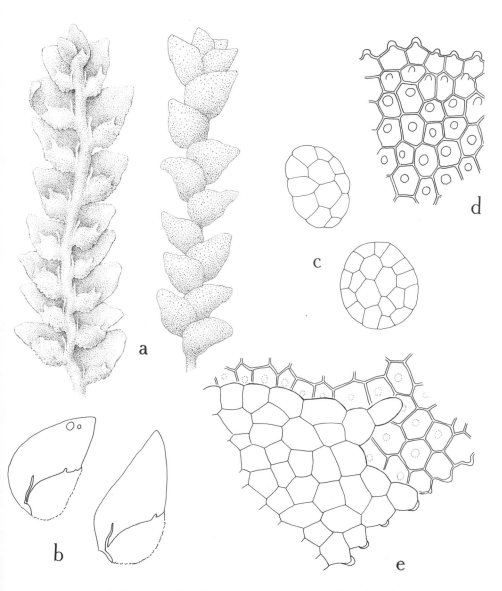

Fig. 96. *Cololejeunea biddlecomiae*. a. Habits, dorsal and ventral views. b. Leaves, one with gemmae on the ventral surface of the larger lobe. c. Gemmae. d. Leaf cells, dorsal view. e. Smooth cells of the ventral lobule adjacent to papillose cells of the dorsal lobe.

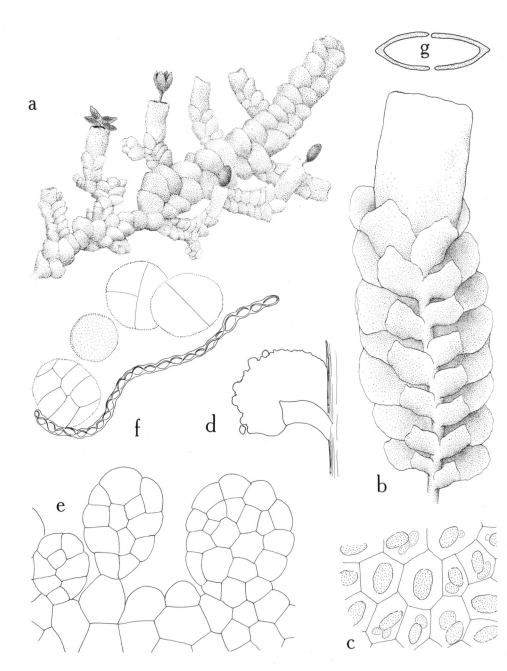

Fig. 97. *Radula complanata*. a. Habit of fruiting plant, dorsal view. b. Habit, ventral view, enlarged. c. Median leaf cells, with oil bodies. d. Leaf that has been eroded by gemma-production, ventral view. e. Gemmae at leaf margin. f. Spores (often germinating precociously within the capsule) and elater. g. Plan of the dorsally compressed perianth formed by the fusion of folded lateral leaves, no underleaf.

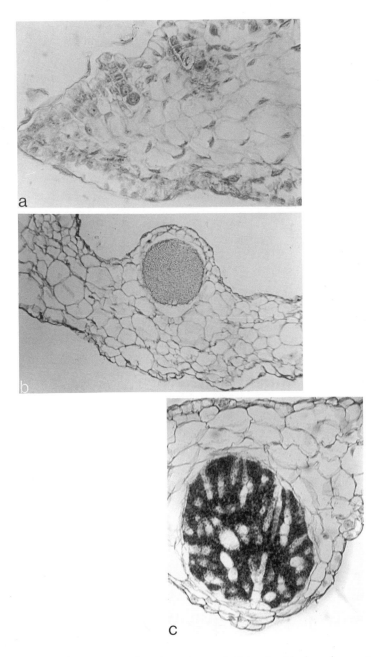

c

Fig. 98. *Anthoceros.* a. Section through 2 embedded archegonia. b. Section through an antheridial cavity with a single antheridium. (Antheridia are more commonly clustered.) c. Ventral slime cavity containing filaments of *Nostoc* (in *A. laevis*).

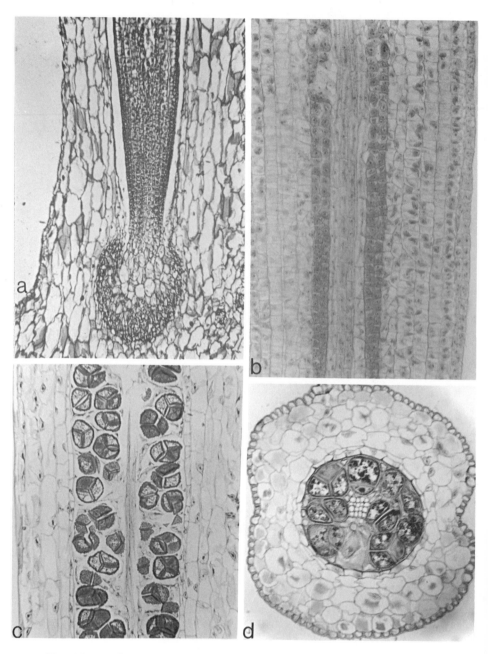

Fig. 99. *Anthoceros.* a. Lower portion of sporophyte showing foot and meristematic tissue, as well as collarlike involucre. b. Section through young portion of sporophyte, showing the central columella surrounded by sporogenous tissue of 2 cell layers and a solid wall. c. Older portion of sporophyte, showing spore tetrads. d. Capsule in section, showing 2 pairs of enlarged guard cells in the epidermis, a capsule wall of several layers of cells in a solid tissue, and spore tetrads surrounding the columella. Some few of the cells not obviously in tetrad groups may represent pseudelaters.

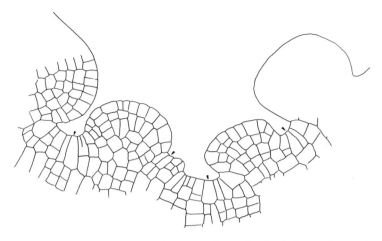

Fig. 100. Dichotomous growth in *Anthoceros fimbriatus* from close-set growing points results in a rounded thallus. (Redrawn from Goebel, *Organography of Plants*, 1905.)

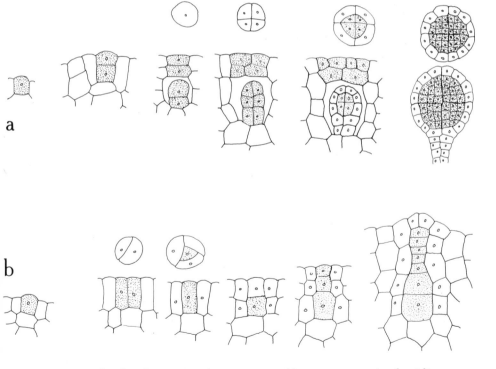

Fig. 101. The development of sex organs of hornworts. a. Antheridia. b. Archegonia. (Redrawn from Schuster, *New Manual of Bryology*, 1984.)

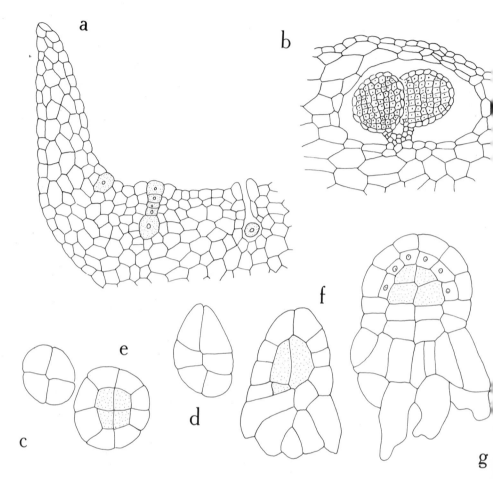

Fig. 102. a–b. Sex organs of *Anthoceros*. a. Portion of thallus in section, showing an archegonial initial and 2 archegonia of different ages. b. Antheridial chamber in section. (Redrawn from Smith et al., *A Textbook of General Botany*, 1928). c–g. Early development of the embryo of *Notothylas orbicularis* (adapted from Campbell, *Mosses and Ferns*). c, d. Quadrant stage in transverse and longitudinal sections. e–g. Differentiation of outer amphithecium from the inner endothecium, as seen in transverse and longitudinal sections. The endothecium gives rise to the columella, and the amphithecium becomes the capsule wall and the sporogenous tissue.

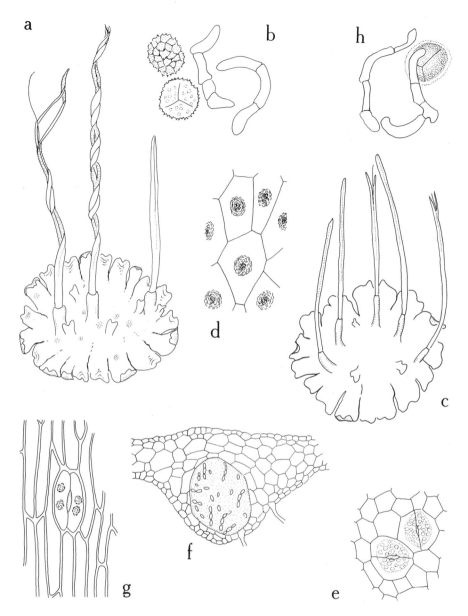

Fig. 103. *Anthoceros punctatus.* a. Habit. b. Spores and pseudelaters. *Anthoceros laevis.* c. Habit. d. Cells of upper epidermis, showing chloroplasts. e. Temporary clefts in lower epidermis near growing point that allow entrance of *Nostoc* to ventral slime cavities. f. Section through thallus, showing *Nostoc*-containing ventral slime cavity. g. Stoma in the epidermal layer of a capsule wall. h. Spore and pseudelaters.

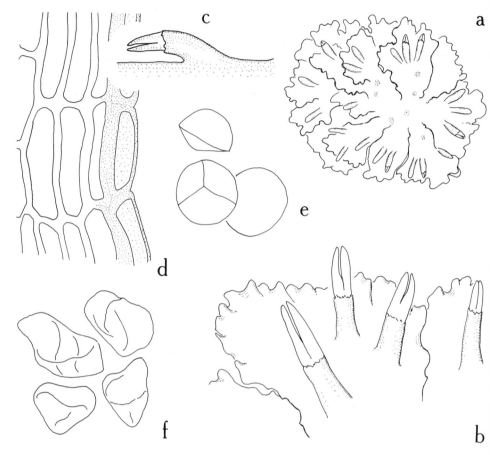

Fig. 104. *Notothylas orbicularis.* a. Thallus. b. Portion of thallus, showing involucre and emergent capsules. c. Involucre and capsule in profile view. d. Epidermal cells of capsule wall along line of dehiscence. e. Spores. f. Pseudelaters, showing mere suggestions of spiral thickenings.

SUBJECT INDEX

TAXONOMIC INDEX

An asterisk (*) indicates an illustration. Italics are used for speeies descriptions.